Evolution and Systematics of the Crassulaceae

Henk 't Hart and Urs Eggli
Editors

Backhuys Publishers
Leiden - 1995

Front-cover illustrations:
Crassula multicava Lem.
Sempervivum armenum Boiss. & Huet (photographs Frits Kindt)

CIP-DATA KONINKLIJKE BIBLIOTHEEK, DEN HAAG

Evolution

Evolution and systematics of the Crassulaceae / Henk 't Hart and Urs Eggli ed. - Leiden : Backhuys Publishers. - Ill.
ISBN 90-73348-46-3
NUGI 824
Subject headings: Crassulaceae ; evolution / Crassulaceae ; systematics

Printed in The Netherlands

Contents

Preface

"A Time of Change" would be a suitable heading to describe the present state of Crassulaceae systematics. The first steps to review and prepare a 'consensus classification' of the Crassulaceae some 50 years after the publication of the last monographic treatment of this family were taken several years back. This initiative culminated in a round-table discussion at the 1992 Congress of the "International Organization for Succulent Plant Study, IOS" in Phoenix (Eggli, IOS Bull. 5. 1992). It is a great pleasure to report now that the original initiative has resulted in much more than a mere 'consensus classification' or 'umbrella' for diverging classificatory views (Eggli & al., this volume). The papers presented during the symposium "Evolution and Systematics of the Crassulaceae" during the 23rd Congress of IOS in Wageningen on August 20th 1994, which are united in this volume, show that significant advances have been made in Crassulaceae systematics in recent years - by 'modern' methods and 'traditional' techniques alike - and we are very grateful to all the participating colleagues for sharing their knowledge. While the 'definite' classification of the family (if such exists) is still a distant goal, the following contributions show a broad agreement in basic concepts and this general mutual understanding provides a firm base for subsequent (and still much-needed) work to improve the ideas now presented.

Henk 't Hart, Urs Eggli
Utrecht, Zürich, March 1, 1995

Contributors

Miguel J. Cházaro Basañez
Faculdad de Geografia, Dep. Biogeografia, Apartado Postal 30, Zapopan, CP 45101 Jalisco, México (Present address: Centro de Botanica, Colegio de Postgraduados, 56230, Chapingo, Estado de Mexico, México)

Urs Eggli
Städtische Sukkulenten-Sammlung, Mythenquai 88, CH-8002 Zürich, Switzerland

Roeland C.H.J. van Ham
Department of Plant Ecology and Evolutionary Biology, Utrecht University, Padua laan 8, 3584 CH Utrecht, The Netherlands (Present address: Department of Genetics, University of Valencia, Dr. Moliner 50, E-46100 Burjassot, Valencia, Spain)

Henk 't Hart
Department of Plant Ecology and Evolutionary Biology, Utrecht University, Padualaan 8, 3584 CH Utrecht, The Netherlands

Norbert Jürgens
Botanisches Institut, Universität Köln, Gyrhofstrasse 15, D-50923 Köln, Germany

Ted H.M. Mes
Department of Plant Ecology and Evolutionary Biology, Utrecht University, Padualaan 8, 3584 CH Utrecht, The Netherlands

Reto Nyffeler
Institute for Systematic Botany, University of Zürich, Zollikerstrasse 107, CH-8008 Zürich, Switzerland

Hideaki Ohba
Department of Botany, University Museum, University of Tokyo, Hongo 7-3-1, Tokyo 113, Japan

Jan F. Stevens
Department of Pharmaceutical Biology, University of Groningen, Ant. Deusinglaan 2, 9713 AW Groningen, The Netherlands

Joachim Thiede
Botanisches Institut der Universität, Olshausenstrasse 40, D-24098 Kiel, Germany (Present address: Botanisches Institut, Universität Köln, Gyrhofstrasse 15, D-50932 Köln, Germany)

CHAPTER 1

Introduction: Evolution of Crassulaceae systematics

Henk 't Hart and Urs Eggli

Introduction

The family Crassulaceae was erected by Augustin Pyramus De Candolle in 1801 and his concept of this group of succulent plants has been a very sound and successful one. De Candolle's delimitation of the Crassulaceae was instantly accepted and has never been seriously challenged, except for the position of the two species of *Penthorum* which have been considered members of the Crassulaceae, the Saxifragaceae, or the monogeneric Penthoraceae (van Tieghem, 1898). In fact recent studies all confirm that the Crassulaceae are a "natural" group. In contrast to the distinctness of this large family (about 1500 species), the infrafamilial classification and the delimitation of a majority of genera of Crassulaceae are subject to an ongoing debate.

Since the conception of the family almost 200 years ago there have been several attempts to describe the total diversity of the Crassulaceae and to present its classification or evolution in a diagram. As an introduction to the present volume of studies on the evolution and systematics of the Crassulaceae an overview of the theories and graphical concepts of the infrafamilial classification and evolution of the family is presented. This summary of hypotheses on the evolution and systematics of the Crassulaceae is first of all meant to provide a historical perspective to the research papers presented in the following chapters. Simultaneously this résumé may indicate the position of modern Crassulaceae research; what has already been achieved and, even more importantly, what still needs to be done.

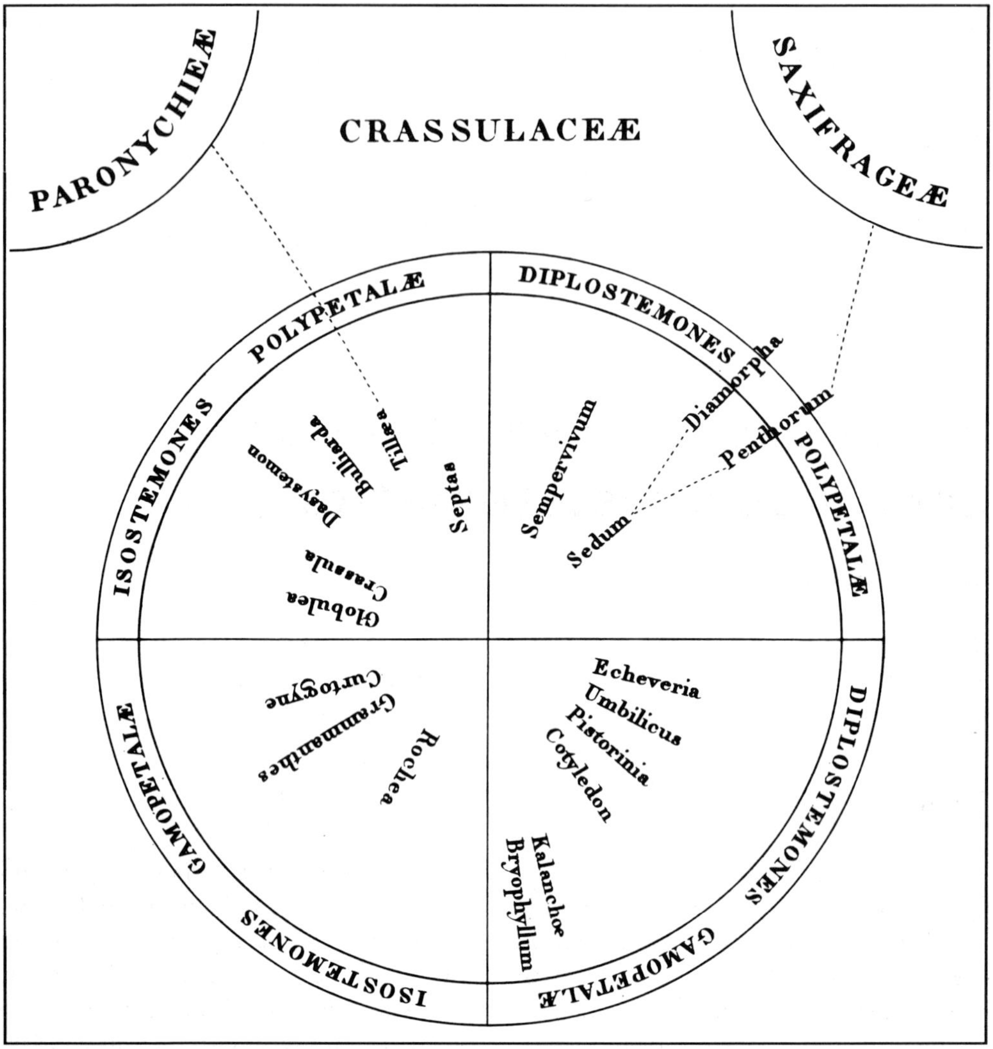

Figure 1. De Candolle's (1828b) synoptical diagram of the classification of the Crassulaceae.

Augustin P. De Candolle

In Prodromus Regni Vegetabilis De Candolle (1828a) listed 19 genera with 308 species of Crassulaceae (including *Penthorum*) about half of which originated from Southern Africa. He divided the family into two tribes: the Crassuleae Legitimae with some 305 species in 17 genera and the Crassulaceae Anomalae which comprised two genera with three species, i.e. *Diamorpha* (1 species) and *Penthorum* (2 species). In contrast to the "true" Crassulaceae which have free carpels, the Crassulaceae Anomalae are characterized by fused carpels. De Candolle (1828b) further developed the classification of the "true" Crassulaceae in a special account of the family. He distinguished four major taxa (Fig. 1) using two floral characters, i.e. the number of stamens (haplostemonous versus obdiplostemonous) and the fusion of the petals (sympetalous versus choripetalous). Because he indicated the number stamens to be the most important of the two characters the Isostemonae and Diplostemonae can be regarded as subtribes, though no formal rank for these taxa is indicated. Both subtribes are divided into two informal groups (no rank being indicated), the Gamopetalae and Polypetalae, respectively.

Selmar Schönland

When Schönland (1891) wrote the account of the Crassulaceae for the first edition of Engler & Prantl's "Die natürlichen Pflanzenfamilien" the number of Crassulaceae species known to science had increased to some 469 species. Schönland's arrangement of the genera is essentially identical to the classification proposed by De Candolle (1828a,b). However, he reduced the number of genera to 13 (reducing many of De Candolle's genera to sections), 5 of which are monotypic. The logic for the reduction of the number of genera is to be found in his complaint about the artificiality of many of the Crassulaceae genera. He emphasized this complaint with some well-chosen examples demonstrating the impossibility to distinguish satisfactorily between *Sedum* and *Crassula*, *Cotyledon*, and *Sempervivum* (including *Aeonium*, *Aichryson*, and *Greenovia*), as well as between *Bryophyllum* and *Kalanchoe* as a result of parallelism of discriminating characters such as polymerous flowers, haplostemony, sympetaly, and synsepaly. Nowadays, however, by far the most interesting aspect of Schönland's account is the graphic presentation of the "evolutionary" relationships among these 13 genera in a diagram (Fig. 2) which, rather surprisingly, closely resembles a modern unrooted Wagner tree. He placed *Sedum* in the centre of the diagram stating that this genus represents the oldest form or "Typus" of the family. This view has been widely accepted and frequently reiterated up till the present (Uhl, 1963; Clausen, 1975; 't Hart, 1982, 1991).

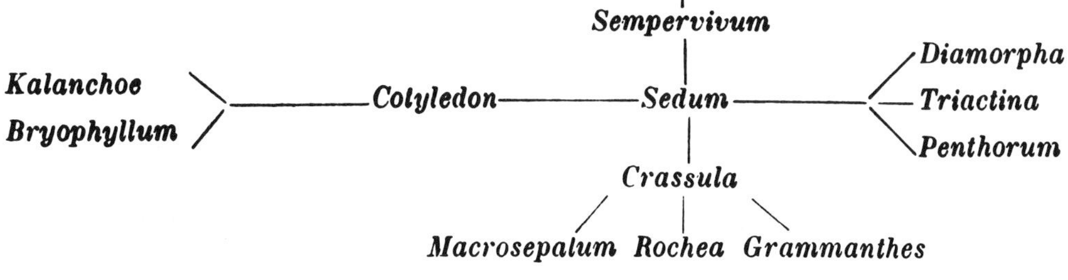

Figure 2. Schönland's (1891) diagram illustrating relationships among the genera of Crassulaceae.

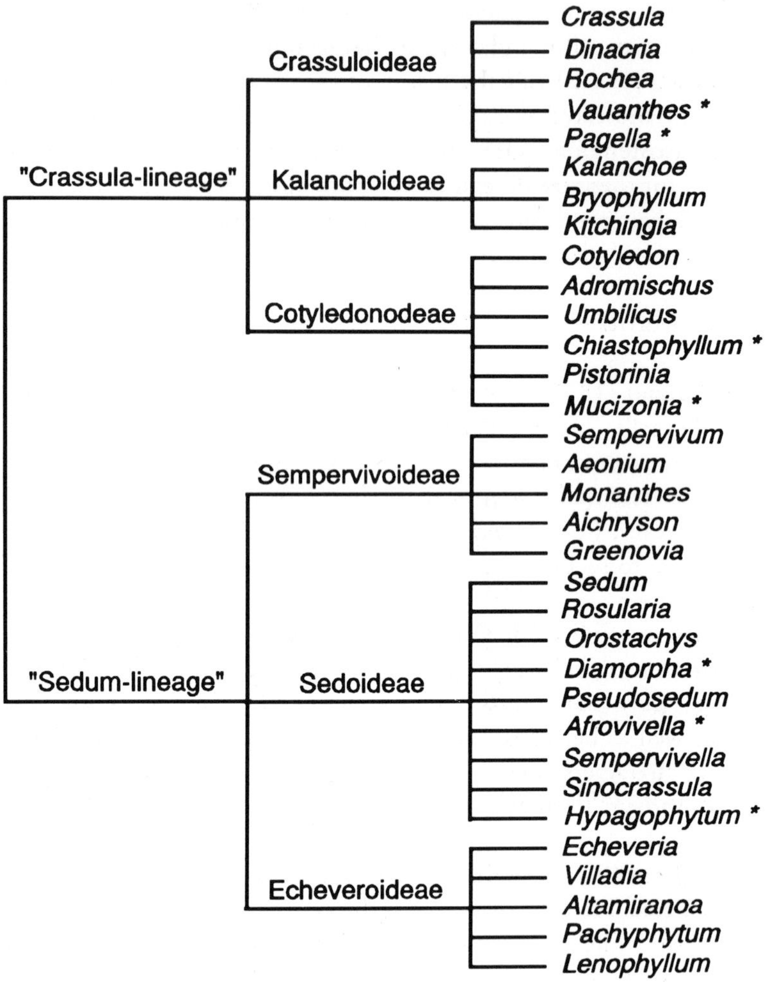

Figure 3. Classification of the Crassulaceae according to Berger (1930). Monotypic genera are indicated with an asterisk.

Alwin Berger

The third major account of the Crassulaceae was written by Berger (1930) for the second edition of Engler & Prantl's "Die natürlichen Pflanzenfamilien". From the very beginning this monograph has been recognized as a great improvement of Crassulaceae systematics and it is still regarded as the most authoritative treatment at the infrafamilial and generic level. It contains about 1400 species in 33 genera including seven monotypic genera (Fig. 3). The genera are arranged in six subfamilies. The five largest genera, i.e. *Sedum, Crassula, Kalanchoe, Echeveria,* and *Sempervivum,* each form the core of the subfamily bearing its name. The six subfamilies are tentatively arranged into two groups of three subfamilies each (Fig. 3), i.e. a *"Crassula"* lineage and a *"Sedum"* lineage, which by and large represent the taxa of the Southern and Northern hemispheres, respectively. The large genera, in particular *Sedum*, are finely subdivided into sections, subsections, series, and subseries. At the infrafamilial as well as at the generic and infrageneric levels Berger applied a traditional and straightforward hierarchical approach to arrange the diversity within this large family. Consequently, the resulting classification is primarily a "practical classification", though as such extremely successful, and almost completely free of evolutionary speculations.

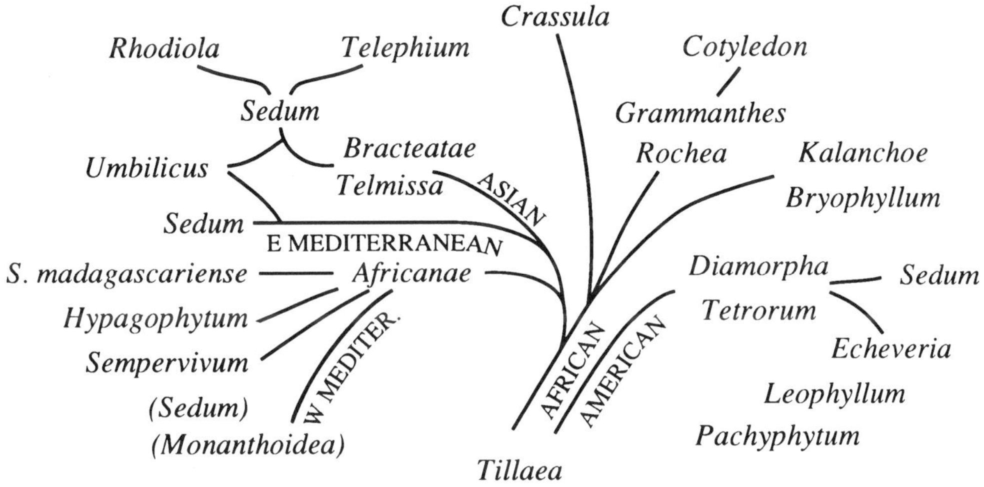

Figure 4. Evolutionary relationships within the Crassulaceae according to Fröderström (1930, 1931).

Harald Fröderström

In the introduction to a monograph of the genus *Sedum* s.l. Fröderström (1930, 1931) developed a completely new and revolutionary theory on the origin and evolution of the Crassulaceae. He rejected Schönland's views about the central position of *Sedum* in the Crassulaceae phylogeny, and in contrast proposed a development from the almost cosmopolitan *Tillaea.* Being very much phytogeographically orientated Fröderström's vivid description of Crassulaceae evolution is full of distributional allegations. Although Fröderström's idea of a "Tillaeoid" origin of the Crassulaceae is generally refuted, as the small and often semi-aquatic *Tillaea* species are highly specialized and adapted to a unique habitat (Uhl, 1963; 't Hart, 1982), his theory also contains some valuable novel hypotheses, in particular with respect to the origin of *Sedum.* Most remarkably, Fröderström considered *Sedum* to be polyphyletic, comprising at least two independent lineages (Fig. 4): 1) A distinct American *Sedum* branch which is the sistergroup of *Echeveria* and related genera. 2) The Asian, European and North African *Sedum* branches wich evolved from a northwards migrating line of East African

Tillaea. The Asian *Sedum* branch in turn gave rise to *S.* sect. *Rhodiola* (≡ *Rhodiola*) and *S.* sect. *Telephium* (≡ *Hylotelephium*), and the East Mediterranean *Sedum* branch to *Rosularia* and *Umbilicus.* The African *Sedum* branch is the sistergroup of *Sempervivum*, which also includes the Macaronesian genera *Aeonium, Aichryson, Greenovia,* and *Monanthes.*

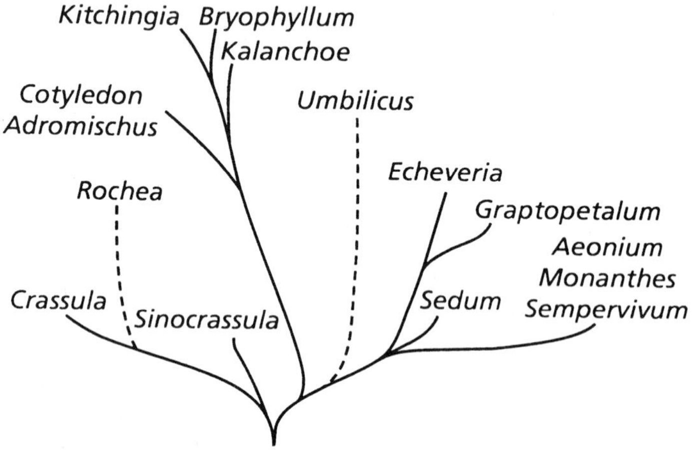

Figure 5. Hideux's phylogeny of the Crassulaceae (after Hideux, 1981).

Michel Hideux

The diagrammatic presentation of the evolution of the Crassulaceae presented by Hideux (1981) is primarily based on the numerical analysis of 58 pollen morphological characters. For the construction of an evolutionary tree, however, he also incorporated cytological characters (basic chromosome numbers), flower morphology, and phytogeographical features in the analysis. The resulting tree (Fig. 5) shows a basal dichotomy, separating a *"Crassula"* branch from all other Crassulaceae. The latter branch splits at the base into an African lineage comprising *Adromischus, Cotyledon,* and *Kalanchoe* (with the subgenera *Bryophyllum* and *Kitchingia*) and a Northern hemisphere lineage which contains *Sedum* and its derivatives, the echeverias (*Echeveria* and *Graptopetalum*) and the sempervivums (*Aeonium, Monanthes,* and *Sempervivum*). The position and affinities of *Rochea* and *Umbilicus* are uncertain. Independently Thorne (1983) proposed a new infrafamilial classification of the Crassulaceae along similar lines. He distinguished only three subfamilies which are fully congruent with Hideux's three major lineages within the Crassulaceae. In addition to the Crassuloideae, he only distinguished the Kalanchoideae (including the Cotyledonoideae), and the Sedoideae (including the Echeverioideae and the Sempervivoideae).

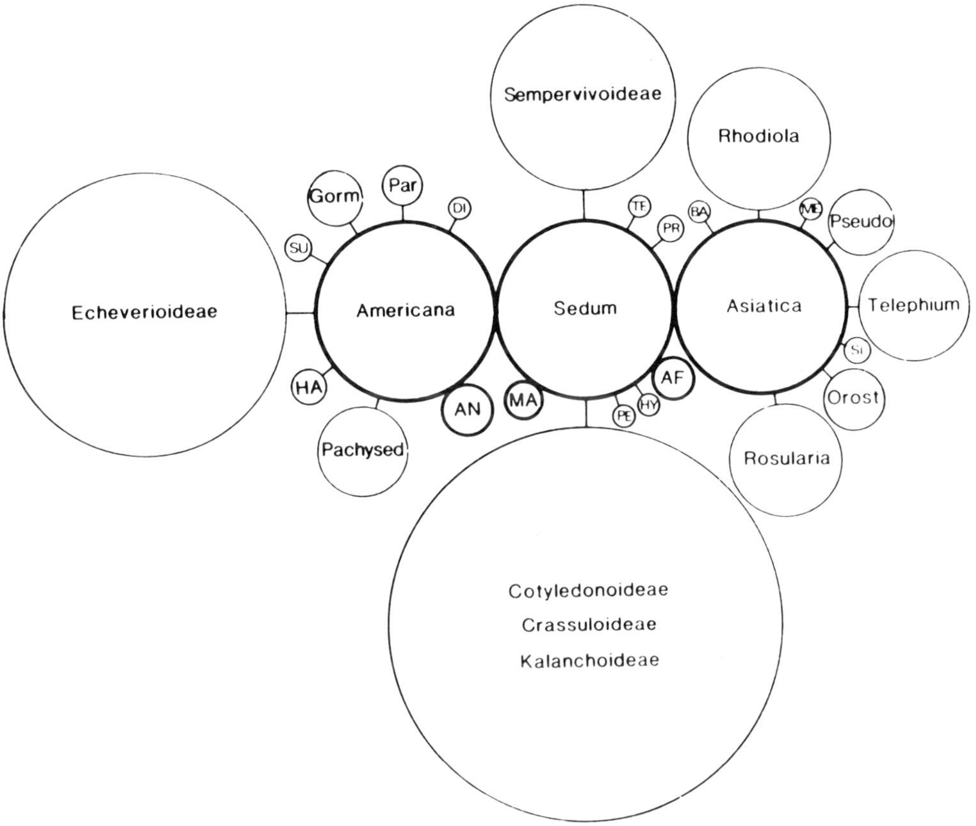

Figure 6. 't Hart's diagram of the classification of the Crassulaceae (after 't Hart, 1982). The size of the circles is proportional to the number species of the taxa they represent. The six contacting circles in the centre of the diagram represent the sections of *Sedum* subgen. *Sedum* (AF - *S.* sect. *Africana*, AN - *S.* sect. *Andina*, MA - *S.* sect. *Macaronesia*). All other infrafamilial taxa are all connected to *Sedum* subgen. *Sedum* by short lines (BA - *Balfouria*, DI - *Diamorpha*, Gorm - *Gormania*, HA - *Hasseanthus*, HY - *Hypagophytum*, ME - *Meterostachys*, Pachysed - *Pachysedum*, Par - *Parvisedum*, PE - *Perrierosedum*, PR - *Prometheum*, Pseudo - *Pseudosedum*, Orost - *Orostachys*, SI - *Sinocrassula*, SU - *Sulcus*, TE - *Telmissa*).

Henk 't Hart

The evolutionary trends within the Crassulaceae as outlined in a diagrammatic infrafamilial classification by 't Hart (1982) are primarily based on biogeographical patterns (Fig. 6). Except for the cosmopolitan genus *Tillaea* and for *Sedum*, which occurs on all three continents of the Northern hemisphere, the genera of Crassulaceae are by and large confined to a single continent or floristic region. Using this geographical pattern as a start the family was divided in a Southern branch comprising Berger's (1930) subfamilies Crassuloideae, Cotyledonoideae, and Kalanchoideae, and a Northern branch which encompasses the rest of the family. The three large, geographically defined sections of *Sedum* hold a central position in the diagram as each section is supposed to have given rise to the genera endemic to that particular (sub)continent or region. However, as 't Hart already (l.c.) clearly stated, it remains to be demonstrated whether these sections are natural taxa, viz. whether it can be shown that they are monophyletic.

Summary

Although these six accounts of the systematic and evolutionary infrafamilial relationships of Crassulaceae appear to be completely different at first sight, it is also apparent that the six authors have used only two criteria to define major lineages within the family. The differences in the resulting classifications are due to variation in the combination and weighting of the different states of these two features. De Candolle (1828a,b), Schönland (1891), and Hideux (1981) situated the main dichotomy in the Crassulaceae between taxa with a single whorl of stamens (haplostemonous) and those with two staminal whorls (obdiplostemonous). In contrast, Berger (1930), Fröderström (1931), and 't Hart (1982) used phytogeographical principles to distinguish the main lineages. A phylogeny of the Crassulaceae based on molecular characters (van Ham, 1994, this volume) indicates that these two criteria are not mutually exclusive, but on the contrary both highlight major developments within the family. All six authors have emphasized the affinities between *Kalanchoe*, *Bryophyllum*, *Cotyledon*, and *Adromischus*, because of the unique combination of two characters, the distinctly sympetalous flowers and a South African distribution, although Schönland (1891), Fröderström (1931), and 't Hart (1982) emphasized the apparent parallellism or convergence in the evolution of sympetaly and the artificiality of taxa based on this character (see also Uhl, 1961, 1963). Berger (1930) and Hideux (1981) regarded *Sedum* s.l. (or the Sedoideae) as a monophyletic taxon. Schönland (1891) and 't Hart (1982), on the other hand, emphasized the paraphyly of the genus, indicating that several major taxa of Crassulaceae have evolved from different lineages of *Sedum*. Fröderström (1931) regarded *Sedum* as a polyphyletic group and, moreover, also presumed that *Sedum* is basal to several other Crassulaceae taxa (paraphyletic).

References

Berger, A. 1930. Crassulaceae. In Engler, A. & Prantl, K. (eds.), Die natürlichen Pflanzenfamilien, ed. 2, **18a**, 352-483. Verlag Wilhelm Englemann, Leipzig.

De Candolle, A.P. 1801. Mémoire sur la famille des joubarbes. *Bull. Soc. Philom.* **49**: 1.

De Candolle, A.P. 1828a. Prodromus naturalis regni vegetabilis. **3**. Treuttel & Wurtz, Paris.

De Candolle, A.P. 1828b. Mémoire sur la famille des Crassulacées. Treuttel & Wurtz, Paris.

Clausen, R.T. 1975. *Sedum* of North America north of the Mexican Plateau. Cornell University Press, Ithaca.

Fröderström, H. 1930. The genus *Sedum* L., 1. *Acta Horti Gothob.* **5**(app.): 1-75.

Fröderström, H. 1931. The genus *Sedum* L., 2. *Acta Horti Gothob.* **6**(app.): 1-111.

Ham, R.C.H.J. van. 1994. Phylogenetic implications of chloroplast DNA variation in the Crassulaceae. Ph.D. Thesis, Utrecht University.

Hart, H. 't. 1982. The white-flowered *Sedum* species, 1. Principles of a phylogenetic classification of the Sedoideae (Crassulaceae) and the position of the white-flowered *Sedum* species. *Proc. Kon. Ned. Akad. Wet., Ser. C,* **85**: 663-675.

Hart, H. 't. 1991. Evolution and classification of the European *Sedum* species (Crassulaceae). *Flora Medit.* **1**: 31-61.

Hideux, M. 1981. Le pollen. Données nouvelles de la microscope électronique et de l'informatique. Agence de Coopération Culturelle et Technique, Paris.

Schönland, S. 1891. Crassulaceae. In Engler, A. & Prantl, K. (eds.), Die natürlichen Pflanzenfamilien, **3, 2a**, 23-38. Verlag Wilhelm Englemann, Leipzig.

Tieghem, P. van. 1898. Sur le genre Penthore considéré comme type d'une famille nouvelle les Penthoracées. *J. Bot.* (Morot) **12**: 150-154.

Thorne, R.F. 1983. Proposed new realignments in the angiosperms. *Nordic J. Bot.* **3**: 75-117.

Uhl, C.H. 1963. Chromosomes and phylogeny of the Crassulaceae. *Cact. Succ. J.* (U.S.) **35**: 80-84.

CHAPTER 2

Phylogenetic relationships in the Crassulaceae inferred from chloroplast DNA variation

Roeland C.H.J. van Ham

Abstract

Evolutionary relationships at the subfamilial level in the Crassulaceae were investigated by phylogenetic analysis of chloroplast DNA (cpDNA) variation. Homologous probes from a complete clone bank of the cpDNA of *Sedum album* were used to study restriction-site variation among 44 species, representing 19 genera and the 6 subfamilies of the Crassulaceae. In addition, insertion/deletion mutations and nucleotide sequence variation in the intergenic spacer between the trnL and trnF genes were studied in 12 species. Maximum overall sequence divergence in the family was approximately 7%. Different tree inference methods applied to the data yielded moderately different topologies. Seven major clades were obtained, designated *Crassula-*, *Kalanchoe-*, *Telephium-*, *Sempervivum-*, *Leucosedum-*, *Aeonium-*, and *Acre-*clade. There was very strong support for a basal division within the family, marked by the highest level of sequence divergence, separating the monophyletic subfamily Crassuloideae from all other taxa. Within the Crassuloideae, the genetic divergence between *Crassula* and *Tillaea* was of the same order of magnitude as among the other major lineages. Four of the six subfamilies recognized by Berger (1930) were indicated to be polyphyletic. These include the Cotyledonoideae, Sempervivoideae, Sedoideae and Echeverioideae. The Kalanchoideae and the South African genera *Cotyledon* and *Adromischus* exhibit low levels of cpDNA sequence divergence relative to one another, suggesting a relatively recent radiation. The genus *Sedum* is indicated to be polyphyletic and seems to comprise sister taxa to most of the other genera of the family. Systematic implications and the pros and cons of the use of the chloroplast genome, in particular noncoding regions, to study evolutionary relationships in the Crassulaceae are discussed.

Introduction

The investigations described in this paper were aimed at providing new and independent characters for phylogenetic inference in the Crassulaceae. To this purpose chloroplast DNA variation was sampled in over 100 species, some 7% of the total species diversity. Two approaches have been employed to assay cpDNA variation among species. The first involved a comparative restriction-site analysis. For the second approach nucleotide sequence and length mutation analysis of noncoding cpDNA regions were employed. A discussion of some of the results of these studies will be preceded by short introductions on the chloroplast genome and the family Crassulaceae, respectively.

The chloroplast genome and its use in phylogenetic studies

Chloroplasts are organelles of endosymbiotic origin. They contain their own genome which consists of a circular DNA molecule. The genome has been extensively studied with respect to its structure, sequence identity, gene expression and evolution since its existence has been demonstrated (Ris & Plaut, 1962). These efforts have culminated in the determination of complete chloroplast DNA sequences of two dicots (*Nicotiana tabacum*, Shinozaki & al., 1986; *Epifagus virginiana*, Wolfe & al., 1992), a monocot (*Oryza sativa*, Hiratsuka & al., 1989), a liverwort (*Marchantia polymorpha*, Ohyama & al., 1986), and an alga (*Euglena gracilis*, Hallick & al., 1993). In addition, large portions of the chloroplast genomes of two angiosperms, a gymnosperm and two algae have been sequenced (Palmer, 1991; Sugiura, 1992; Reith & Munholland, 1993) as well as a large number of genes from widely divergent plant species (e.g., Clegg, 1993). Finally, well over one thousand chloroplast genomes have been structurally characterized by physical and gene mapping studies (reviewed in Palmer & al., 1988; Palmer, 1991; Downie & Palmer, 1992). The phylogenetic diversity documented by this large body of data is unparalleled by studies of any other genome.

The vast majority of land plant chloroplast genomes is between 130 and 160 kilo base pairs, which is a rather narrow range compared to plant mitochondrial or nuclear genomes. An conspicuous architectural feature of its overall structure is the presence of a large inverted duplication, the inverted repeat (IR), which divides the molecule asymmetrically into two single copy regions. The most important feature of cpDNA is that it is very densely packed with genetic information, with intergenic spacers of, on average, only a few hundred base pairs long. The vast majority of the approximately 115 genes identified on land plant chloroplast genomes function in the processes of gene expression and photosynthesis. Many genes are organized into operons and several of these are still very similar to their eubacterial counterparts. Comparisons of synonymous substitution rates in protein coding genes from different eukaryotic genomes reveal that angiosperm cpDNA evolves relatively slowly (Wolfe & al., 1987, 1989; Wolfe & Sharp, 1988; Palmer, 1991). Only angiosperm mitochondrial genes evolve more slowly, at an approximately three-fold lower rate. In contrast, angiosperm and mammalian nuclear genes evolve approximately two to five times faster, and mammalian mitochondrial genes even up to forty times faster than cpDNA genes. Remarkably, genes in the chloroplast IR evolve about five times slower than genes in the single copy regions, suggesting that the IR is under strong functional constraints (Palmer, 1991).

In general the land plant chloroplast genome is the most conservatively evolving eukaryotic genome. Apart from some great practical advantages of cpDNA, including its abundance in plant cells, the ease with which it can be isolated and its manageable size, the overall conservative mode of evolution implies that whenever changes in any of its features occur, these can be of great value for the inference of evolutionary relationships. Despite its evolutionary conservativeness, there is a window of variation that covers all taxonomic levels, even the intraspecific level (reviewed in Soltis & al., 1992; Olmstead & Palmer, 1994). For phylogenetic purposes, it is the taxonomic level, or better, the amount of evolutionary divergence observed or expected in the group under study, which determines the technical approach as well as the particular aspect of cpDNA variation that can best be studied. Some generalizations about this relation can be made.

From the intraspecific up to approximately the familial level, noncoding sequences or rapidly evolving gene sequences, restriction-site variation, or small length mutations can be fruitfully studied. The most widely employed approach in studies of plant phylogeny at this level, however, is comparative restriction-site analysis (Palmer & al., 1988). Most of these studies have focused on relationships between congeneric species and closely related genera. The number of studies at higher infrafamilial levels is still rather limited.

Above the familial level, the aforementioned approaches usually fail because the levels of overall sequence divergence and small length variation are usually too high. Here, the approach can be to determine nucleotide sequences of genes with an optimal fit of variation relative to the evolutionary divergence of the group under study, or to search for major structural rearrangements such as inversions, gene and intron gains/losses, or loss of the IR. Relatively speaking, inversions occur most frequently during chloroplast genome evolution. Because of the very low probability with which inversions are expected to occur independently more than once at the same position within a genome, such mutations are extremely powerful markers of phylogenetic relationships.

The Crassulaceae

The Crassulaceae is generally considered a distinctive and monophyletic family among angiosperms (Berger, 1930; Cronquist, 1981; Thorne, 1983; Haskins & Hayden, 1987). It is characterized, first, by a relatively primitive floral structure (Takhtajan, 1980), consisting of regular, usually 5-merous flowers, with isomerous, free carpels and one or two whorls of stamens. Secondly, the family exhibits a number of xerophytic features that reflect its adaptation to arid habitats. These include, foremost, leaf succulence, a thick, waxy cuticle, an often rosulate habit, and tufted or matted growth-forms (Denton, 1982). Generally associated with these traits (Pilon-Smits, 1992) is the physiological adaptation to which the family lent its name, i.e., Crassulacean acid metabolism (CAM).

The family is of nearly cosmopolitan distribution (except Polynesia), but is very rare in Australia and South America. Crassulaceae are most common in (semi-)arid and mountainous habitats of the temperate and subtropical zones, with conspicuous centres of diversity occurring in Mexico, South Africa, and the Mediterranean and Himalayan regions. The Crassulaceae are generally classified in the Rosalean complex, within which they are thought to be most closely related to Penthoraceae and Saxifragaceae (Takhtajan, 1980; Cronquist, 1981; Thorne, 1983). Recent molecular data pertaining to serological characters (Grund & Jensen, 1981) and DNA sequence comparisons of the chloroplast rbcL gene (Albert & al., 1992a) have confirmed these affiliations.

Biologically, Crassulaceae are most exceptional by a syndrome of extreme cytological variation, a widely retained crossability, and extensive homoplastic evolution of morphological features (Berger 1930; Fröderström 1930, 1932; Mauritzon, 1933; Baldwin, 1939; Quimby, 1939; Uhl, 1948, 1961a, 1963, 1992; Wassmer, 1955; Jensen, 1968; Clausen, 1975; Ohba, 1978; 't Hart, 1982, 1985, 1991; 't Hart & Koek-Noorman, 1989; Grulich, 1984; Eggli, 1988). An outstanding example of the first two phenomena is the interconnection by artificial hybrids of about 200 species in and around the genus *Echeveria*. This group represents one of the largest comparia known, which in addition covers a continuous series of basic chromosome numbers of x=12 to x=34, with higher numbers up to n=260 (Uhl, 1992). The most conspicuous examples of homoplastic developments in morphological characters, which have frequently been used for infrafamilial classification, include: annual and rosulate habit; monopodial axis; subterranean, fleshy caudex; secondary woodiness; decussate and whorled phyllotaxis with an accompanying closed primary vascular pattern; presence of glandular hairs; reduction and/or increase of the number of floral parts; fusion of sepals; fusion of petals and formation of a long corolla tube; number of staminal whorls; position and shape of the ripe follicles; ornamentation of the testa. Some of these morphological characters may be under simple genetic control, which could explain the high frequency with which they evolved, either by chance or under similar selective pressures, in independent lineages within the family.

Of the three phenomena mentioned, homoplastic character evolution most obviously underlies the notoriously difficult delimitation of many infrafamilial taxa of the Crassulaceae. This circumstance is further aggravated by a general paucity of distinct morphological developments in the family. A complaint uttered by Linnaeus has been reiterated in nearly every systematic treatment of the group: "the genera resemble each other so closely that the limits between them cannot be fixed unambiguously" (Fröderström, 1930). As a result, various infrafamilial classifications have been proposed (reviewed in 't Hart, 1982) but none of these has ever been unanimously supported by taxonomists.

Berger (1930) recognized 6 subfamilies and 33 genera. The subfamilies are based on the number and arrangement of floral parts, the degree of sympetaly, and the phyllotaxis. Each of them comprises one of the larger genera *Crassula*, *Kalanchoe*, *Cotyledon*,

Sempervivum, *Sedum*, and *Echeveria*. In addition, Berger tentatively circumscribed two major groups of three subfamilies each; the "*Crassula*-lineage" comprises the Crassuloideae, Kalanchoideae, and Cotyledonoideae, and the "*Sedum*-lineage" comprises the Sempervivoideae, Sedoideae, and Echeverioideae. The former group has a predominantly South-African distribution, while the latter is mainly confined to the Northern Hemisphere. Thorne (1983) proposed a reduction of the number of subfamilies to three, recognizing only the Crassuloideae, Cotyledonoideae (including Kalanchoeae) and Sedoideae (including Sempervivieae and Echeveriaea).

Owing to its comprehensiveness and great practical value, the classification proposed by Berger (1930) has been authoritative and most widely followed, despite the general conviction that a substantial part of it is highly artificial (Fröderström, 1932; Uhl, 1961a, 1963; Clausen, 1975; Ohba, 1978; 't Hart, 1982, 1991; Grulich, 1984; Eggli, 1988). The core of the systematic problems in the Crassulaceae lies in the delimitation and infrageneric classification of the largest genus *Sedum*, which with about 500 species, accounts for about one third of the family's total number of species. Although only few of the other Crassulacean genera can be defined by unique features, there is still considerable consensus on the delimitation and naturalness of most of those recognized by Berger (Eggli, 't Hart & Nyffeler, this volume). In contrast, *Sedum* is generally considered as a dustbin-taxon that encompasses the least derived Crassulaceae as well as homoplastic or transitional phenotypes to most of the other genera of the family. Consequently, crossability as well as biogeographical and cytological criteria have played a major role in systematic treatments of *Sedum* (Fröderström, 1930; Uhl, 1961a,b, 1963; Clausen, 1959, 1975; Ohba, 1978; 't Hart, 1978, 1991).

The central position attributed to *Sedum* in the evolution of the Crassulaceae is visualized in the diagrammatic classification of the family proposed by 't Hart ('t Hart & Eggli, this volume Fig. 6), which is based primarily on biogeographical patterns. Except for *Sedum* and the cosmopolitic genus *Tillaea* (subfamily Crassuloideae), the genera of the Crassulaceae are largely confined to a single continent or floristic region. Accordingly, in this classification, the subfamilies Echeverioideae and Sempervivoideae, the genera reckoned to the subfamily Sedoideae, and some infrageneric taxa of *Sedum* are assumed to be most closely related to one of the three large, paraphyletic, and biogeographically defined sections of *Sedum* subgenus *Sedum*.

CpDNA phylogeny of the Crassulaceae

We started with the construction of a clone bank and restriction-site map of the cpDNA of one species, *Sedum album*, with the aid of heterologous probes from lettuce and petunia (van Ham & al., 1992). This resulted in a complete clone bank and restriction-site map (Fig. 1), which is colinear with the consensus angiosperm gene order first found in tobacco (Shinozaki & al., 1986). Clones from this bank were used in subsequent large scale southern hybridization experiments to construct restriction maps for other species. The analysis involved over 100 species. We constructed restriction-site maps for these genomes, using 8-12 rather frequently cutting restriction en-

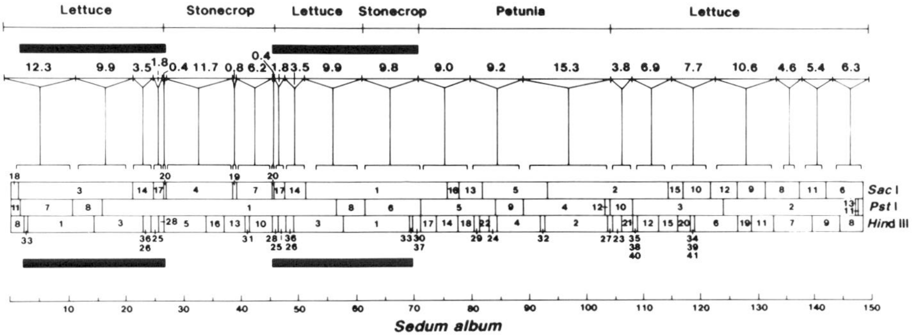

Figure 1. Linear map of restriction endonuclease cleavage-sites of *Sedum album* chloroplast DNA. Positions of cleavage sites generated by the restriction enzymes SacI, PstI and HindIII are given relative to the position of the lettuce and petunia clones used in the heterologous filter hybridizations. Vertical lines from the clones to fragments of *S. album* indicate hybridization results. The inverted repeat regions are indicated by the solid bars below the map.

zymes. Sites were aligned to assess their homology and, apart from a 6 kb sequence located in the large single copy region, which gave highly complex hybridization signals, the maps showed no rearrangements relative to each other. All genomes proved to be about 150 kb in size, with the largest deletion of 1 kb found in a single species. This procedure resulted in some 1200 variable sites for the whole group of species, implying some 130,000 datapoints to be used in phylogenetic analysis. Phylogenetic analysis of the data was performed with PAUP version 3.1.1 (Swofford, 1993) and PHYLIP version 3.4 (Felsenstein, 1991). The Neighbor-joining (NJ) method was used to construct an unrooted distance network based on overall cpDNA sequence divergence values. Parsimony analyses involved consideration of the Wagner, Dollo and generalized parsimony criteria (Swofford & Olsen, 1990), for the latter using the weighting-scheme for restriction-site changes proposed by Albert & al. (1992b). Bootstrap analysis (Felsenstein, 1985) and decay analysis (Donoghue & al., 1992; Soltis & al., 1993) were performed to examine the support for branches in most parsimonious trees.

The main purpose of our investigations was to identify, at least molecularly, major groups within the family and their interrelationships. Here, we concentrate on a subset of 46 species that for the total group of species represents the major lineages within the family. Maximum sequence divergence along the basal dichotomy in the family was approximately 7.3 %. Upon 100 random-taxon-addition tree searches with options of steepest descent and TBR-branch swapping, Wagner parsimony analysis of these 46 species resulted in single most parsimonious tree (Fig. 2). The tree has a length of 1782 steps and a consistency index (CI) of 0.34. The number of mutational steps, and bootstrap and decay values are indicated along the branches. We distinguish seven major clades in this tree that almost always turned up in our phylogenetic analysis using different methods or sampling of species. The clades are indicated at the right of the species names in Figure 2:

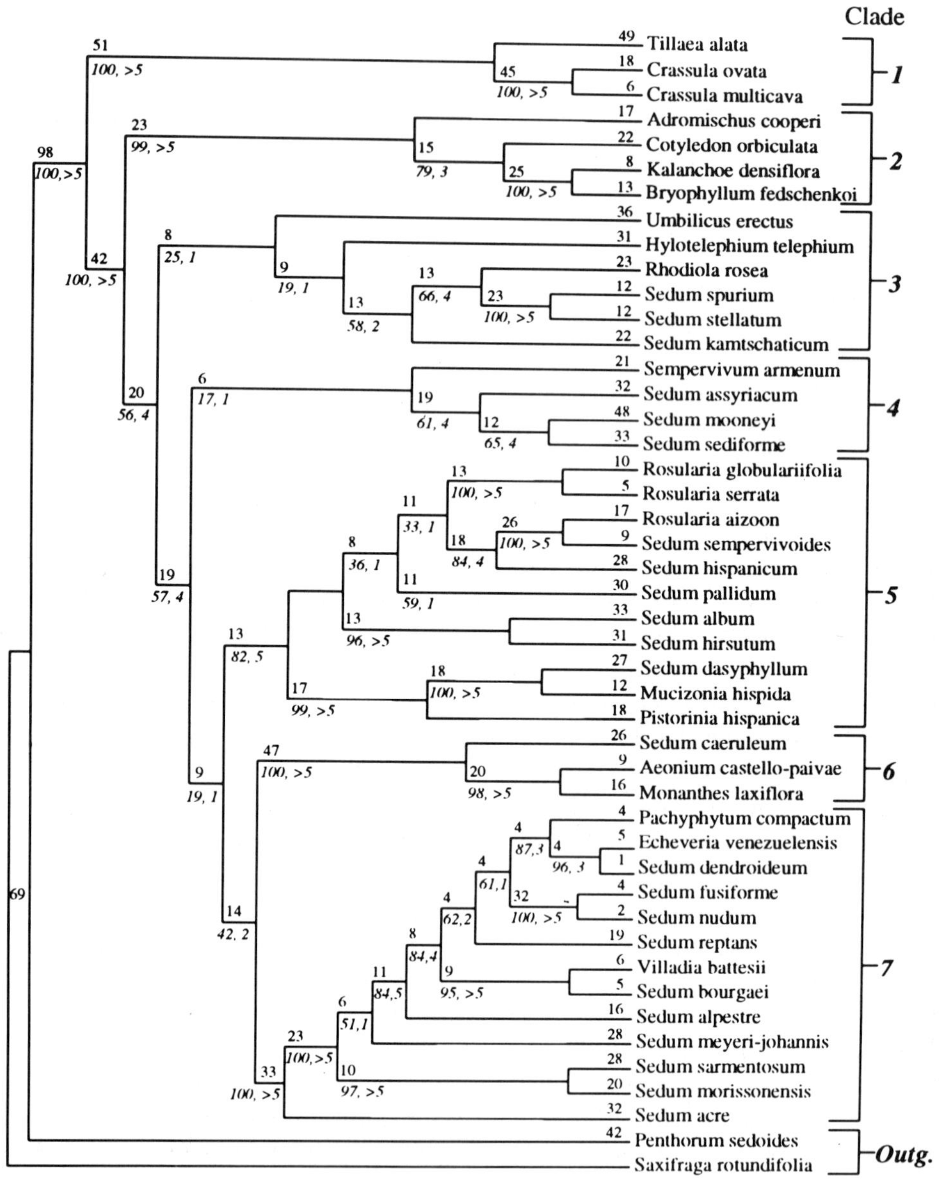

Figure 2. Single most parsimonious Wagner tree of Crassulaceae, using 608 phylogenetically informative cpDNA restriction-sites. The 50% majority rule consensus tree obtained from a bootstrap analysis, performing 500 replicates, was fully compatible with this tree. The tree has a length of 1782 steps and a consistency index of 0.341 (0.425 including autapomorphies). Branch lengths are shown above the lines. Bootstrap and decay values, respectively, are shown below the lines at each node. Seven clades considered to be of major importance in the discussion of the evolution of the family are indicated by brackets and designated by numbers.

1) The first is the *Crassula*-clade, which is strongly supported by our data and comprises only members of Berger's subfamily Crassuloideae. It is characterized by haplostemonous flowers, which occur only extremely rarely in the other 6 clades. It comprises about 300 species and is largely restricted to South Africa. This basal dichotomy in the family, between Berger's Crassuloideae and the rest of the family, is marked by the highest value of overall sequence divergence, approximately 7.3%, while maximum sequence divergence in the other group is about 5%, which is in the order of size of the maximum sequence divergence observed in the large family Compositae by Jansen & Palmer (1988). An interesting aspect of this group is the high value of sequence divergence between *Crassula* and the disputed genus *Tillaea*, which Berger classified as a section of *Crassula*. The estimated value of about 5% is in the range of the maximum divergence observed in the bulk of the second major group of Crassulaceae.
2) The second major and also strongly supported clade, which represents about 250 species, comprises Berger's subfamily Kalanchoideae and the South African genera of the Cotyledonoideae. Sequence divergence in the *Kalanchoe*-clade is surprisingly low with a maximum value of about 2%. This value is close to the sequence divergence we observed in a small comparium of 7 European species, *Sedum* series *Rupestria* (van Ham & 't Hart, 1994), suggesting that the *Kalanchoe* group has diversified relatively recently. The three Mediterranean genera of the Cotyledonoideae (*Umbilicus*, *Pistorinia*, and *Mucizonia*) appear in different clades in the family tree, indicating that the Cotyledonoideae in Berger's circumscription are polyphyletic.
3) One of the Eurasian genera of the Cotyledonoideae, *Umbilicus*, is basal to the third major clade we distinguish in this tree, the *Telephium*-clade after *Hylotelephium telephium*. Taxonomically this is a heterogeneous group and not so strongly supported at its base. It unites species of the Northern hemisphere with flat and dentate leaves, a feature which is otherwise restricted to members of the first to clades. It is also the first clade in which *Sedum* species show up in this tree.
4) Clade four or *Sempervivum*-clade is also heterogeneous with a single genus from Berger's subfamily Sempervivoideae together with *Sedum* species.
5) Clade 5, the *Leucosedum*-clade named after *Sedum album*, comprises the first large group of *Sedum* species together *Rosularia* which is polyphyletic in this tree, and two other Mediterranean genera of the Cotyledonoideae, *Mucizonia* and *Pistorinia*.
6) Clade 6, the *Aeonium*-clade, comprises the Macaronesian Sempervivoideae (*Aeonium* and *Monanthes*) together with *Sedum* species. It is a strongly supported and molecularly highly differentiated group, and gives an indication that, after the Cotyledonoideae and Sedoideae, the Sempervivoideae are also polyphyletic. Sequence divergence between *Aeonium* and *Monanthes* is less than 1% which may confirm earlier suggestions of recent adaptive radiation in this group.
7) Finally, there is a large and strongly supported clade, the *Acre*-clade, which comprises members of the strictly American Echeverioideae together with *Sedum* species from all over the world. The Echeverioideae, here represented by *Pachyphytum*, *Echeveria*, and *Villadia* are again also indicated to be polyphyletic. Clade 7 is one of the most surprising groups found, because as a group which shows up in the upper part of the tree, it unites the American, highly differentiated allies of *Echeveria* with what has been hy-

pothesized as the most primitive, or least derived Crassulaceae, viz. *Sedum acre* ('t Hart & Koek-Noorman, 1989; 't Hart, 1992).

We employed some different methods of phylogenetic analysis to assess the robustness of the groupings here presented and the alternatives are depicted in Figure 3. The major outline of these trees, however, was not influenced by taxonomic sampling among the larger group of species we assayed for cpDNA variation. Figure 3A summarizes the Wagner tree (Fig. 2) and the major clades are designated here by their provisional names. Figure 3B summarizes the strict consensus tree of six most parsimonious trees obtained after Dollo parsimony analysis of the data. The *Sempervivum*-clade is broken up, while the other 6 clades remain intact, showing only few internal topological differences. More important in the present context is that there are a number of differences with the Wagner tree with respect to relationships among the major groups. Most striking of these are the grouping of the *Kalanchoe*- and *Telephium*-clades and the placement of the *Acre*-clade basal to the *Sempervivum*-, *Aeonium*-, and *Leucosedum*-clades. Figure 3C summarizes the NJ tree constructed from nucleotide sequence divergence values. It mirrors the relationships between clades 4 to 7 found in the Wagner tree.

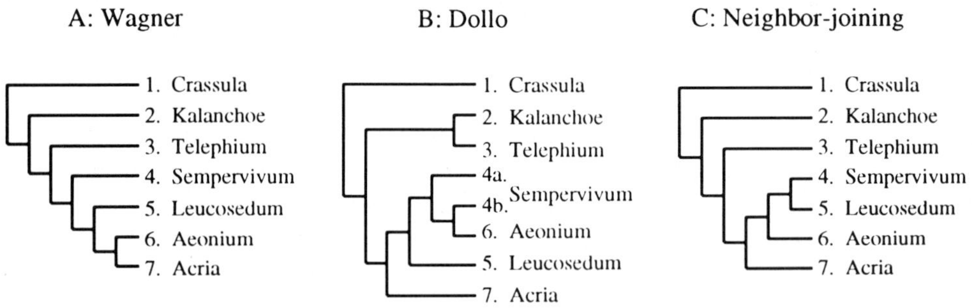

Figure 3. Summary of alternative cpDNA phylogenies obtained by employing different methods of phylogenetic analysis. Only the seven major clades obtained in the Wagner parsimony analysis (Fig. 3) are compared.

With respect to parsimony methods both Wagner and Dollo parsimony are suboptimal for the analysis of restriction-sites (Albert & al., 1992b). Restriction-sites have asymmetric probabilities of being lost or gained, the chance of loosing a particular site being larger than of gaining that same site. In Wagner parsimony, both transformations are weighted equally, whereas in Dollo parsimony one transformation is weighted infinitely over the other. This problem with restriction-sites has long been recognized and generalized or relaxed Dollo parsimony has been proposed as a way out of this problem (Swofford & Olsen, 1990). Albert & al. (1992b) have proposed a model of transformational weights that can be assigned to restriction-site mutations, reckoning with their different probabilities of occurring. However, for our datamatrix, the application of generalized parsimony was computationally too time consuming and we were only able to apply it to a strict consensus tree that we constructed from the Wagner, Dollo, and NJ trees. This approach facilitates the searches under generalized parsimony because it reduces the time-consuming rearrangements that have to be done during a search. Weight ratios for restriction-site gains over losses of 1.1 to 2.3, which is in the range recommended by Albert & al. (1992b), over the full range yielded trees that were highly similar to the Wagner tree of Figure 3A. Even at the upper bound of 2.3 there were no differences with respect to the relationships among the seven major clades. At weight ratios increasingly higher than 3 the topologies gradually converged into the Dollo tree (Fig. 3B). We may conclude from these analyses that, although we could identify at least seven major lineages in the Crassulaceae, the relationships in the group comprising clades 4 to 7 are still not resolved with great confidence. With respect to taxonomic sampling, the grouping of the *Acre-* and *Aeonium*-clade might be due to what has been called long branch attraction and from which parsimony methods suffer in cases were there are highly differentiated groups amid much less differentiated ones (Felsenstein, 1978). Sampling of more obscure taxa might help in this respect to break up eventually some of these long branches and thus help to resolve their interrelationships. Nevertheless, we believe that this restriction-site analysis in the Crassulaceae has been very fruitful in identifying major groups in a bottom-up approach and which may guide future molecular systematic studies in the group. The taxonomic implications are considerable: not less than four of the six subfamilies of Berger, a classification that is generally regarded as authoritative, are indicated to be unnatural assemblages. Most problematic still remains the large genus *Sedum* that is scattered all over the cpDNA phylogeny.

Noncoding regions; the use of small indels as taxonomic markers

In addition to the restriction-site analysis, we investigated cpDNA variation in three adjacent noncoding regions of the genome (Fig. 4), which were amplified with primer sets based on the highly conserved tRNA genes (Taberlet & al., 1991). We observed small overall length differences of PCR-products of a short region of the genome that carries three tRNA genes (van Ham & al., 1994). This region is located in the large small single copy region close to the rbcL gene.

In the intergenic spacer between the trnL and trnF genes there was considerable length variation in the Crassulaceae as well as among the outgroup species. To study the underlying nature of these length differences we sequenced this intergenic spacer in three outgroup species and 12 Crassulaceae that cover a broad range of taxonomic levels. The multiple alignment required a large number of small insertions and deletions. Many insertions proved to be short direct repeats (Zurawski & al., 1984), which are caused by slippage of the DNA polymerase during replication and repair. One of the largest deletions, 136 bp in one of the outgroup species appeared to contain an inverted repeat motif around the deletion, suggesting that it might have occurred by a recombinational process. In total we observed 50 indels among 16 species including tobacco. Of these, 28 are presumably deletions and 22 are clearly duplications or insertions. We wondered whether these small length mutations can be used as reliable markers for evolutionary groupings at lower taxonomic levels similar to the way in which large structural mutations in cpDNA have proven to be powerful markers of relationships at higher taxonomic levels, e.g., a 22 kb inversion found in the Compositae by Jansen & Palmer (1987). To answer this question we need to get insight in the level of homoplasy that underlies these mutations, so first we need a reliable phylogeny. We constructed a tree from the sequence data as well as from the restriction-site data that we sampled for these species and we evaluated the level of homoplasy for these small indels. Interestingly, although there are some differences between the sequence and restriction-site trees, only length mutations smaller than three base pairs required multiple origins over both topologies. Only one mutation, a duplication of four bp located in a relative hot spot of both length and point mutations required two independent origins over both trees. From these results we have an indication that indeed small length mutations larger than about three base pairs can be used as powerful markers of evolutionary relationships at lower taxonomic levels in a similar way as large rearrangements in cpDNA are markers for evolutionary relationships at the higher taxonomic levels.

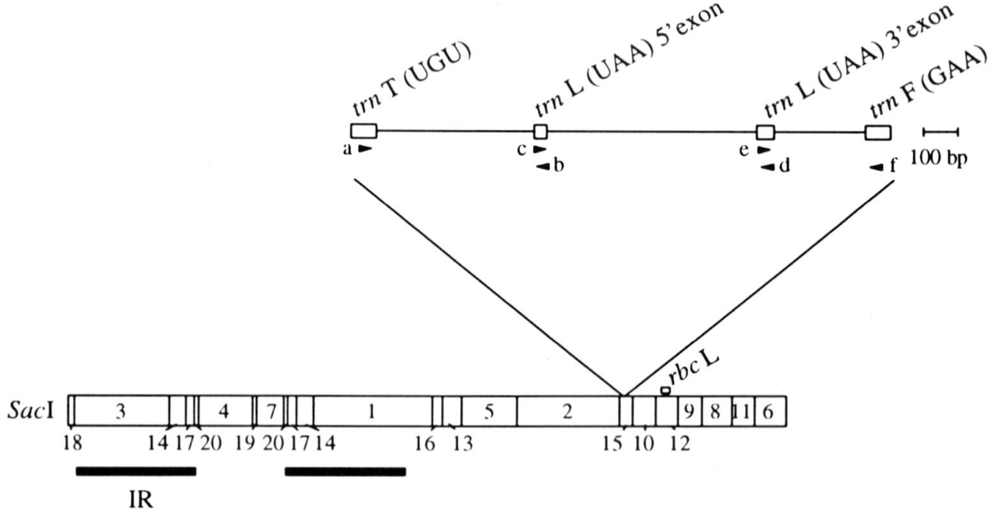

Figure 4. Location of the three noncoding regions on the SacI restriction-site map of *Sedum album* chloroplast DNA (van Ham & al., 1992) as determined by Southern hybridization. Primers used for PCR (Taberlet & al., 1991) are indicatied by arrowheads (a-f).

References

Albert, V.A., Mishler, B.D. & Chase, M.W. 1992a. Character-state weighting for restriction-site data in phylogenetic reconstruction, with an example from chloroplast DNA. In Soltis, P.S., Soltis, D.E. & Doyle, J.J. (eds.), Molecular Systematics of Plants, 369-403. Chapman and Hall, New York.

Albert, V.A., Williams, S.E. & Chase, M.W., 1992b. Carnivorous plants: phylogeny and structural evolution. *Science* **257**: 1491-1495.

Baldwin, J.T. 1939. Certain cytophyletic relations in the Crassulaceae. *Chronica Bot.* **5**: 415-417.

Berger, A., 1930: Crassulaceae. In Engler, A. & Prantl, K. (eds.), Die Natürlichen Pflanzenfamilien, ed. 2, **18a**, 352-483. Verlag Wilhelm Engelmann, Leipzig.

Clausen, R.T. 1959. Sedum of the Trans Mexican Volcanic Belt. Ithaca, New York.

Clausen, R.T. 1975. Sedum of North America north of the Mexican Plateau. Ithaca, New York.

Clegg, M.T. 1993. Chloroplast gene sequences and the study of plant evolution. *Proc. Natl. Acad. Sci. USA* **90**: 363-367.

Cronquist, A. 1981. An integrated system of classification of flowering plants. Columbia University Press, New York.

Denton, M.F. 1982. Revision of *Sedum* section *Gormania* (Crassulaceae). *Brittonia* **34**: 48-77.

Donoghue, M.J., Olmstead, R.G., Smith, J.F. & Palmer, J.D. 1992. Phylogenetic relationships of Dipsacales based on rbcL sequences. *Ann. Missouri Bot. Gard.* **79**: 333-345.

Downie, S.R. & Palmer, J.D. 1992. Use of chloroplast DNA rearrangements in reconstructing plant phylogeny. In Soltis P.S., Soltis, D.E. & Doyle, J.J. (eds.), Molecular Systematics of Plants, 14-35. Chapman and Hall, New York.

Eggli, U. 1988. A monographic study of the genus *Rosularia* (Crassulaceae). *Bradleya* **6**(suppl.): 1-119.

Felsenstein, J. 1978. Cases in which parsimony or compatibility methods will be positively misleading. *Syst. Zool.* **27**: 401-410.

Felsenstein, J. 1985. Confidence limits on phylogenies: an approach using the bootstrap. *Evolution* **39**: 783-791.

Felsenstein, J. 1991. PHYLIP: Phylogeny Inference Package, Version 3.4., Computer program distributed by the University of Washington, Seattle, Washington.

Fröderström, H. 1930. The genus *Sedum* L. 1. *Acta Horti Gothob.* **5**(app.): 1-75.

Fröderström, H. 1932. The genus *Sedum* L. 3. *Acta Horti Gothob.* 7(app.): 1-125.

Grulich, V. 1984. Generic division of Sedoideae in Europe and adjacent regions. *Preslia* **56**: 29-54.

Grund, C. & Jensen, U. 1981. Systematic relationships of the Saxifragales revealed by serological characteristics of seed proteins. *Pl. Syst. Evol.* **137**: 1-22.

Hallick, R.B., Hong, L., Drager, R.G., Favreau, M.R., Monfort, A., Orsat, B., Spielmann, A. & Stutz, E. 1993. Complete sequence of *Euglena gracilis* chloroplast DNA. *Nucleic Acids Res.* **21**: 3537-3544.

Ham, R.C.H.J. van & Hart, H. 't. 1994. Evolution of *Sedum* series *Rupestria* (Crassulaceae): evidence from chloroplast DNA and biosystematic data. *Pl. Syst. Evol.* **190**: 1-20.

Ham, R.C.H.J. van, Mes, T.H.M., Hart, H. 't & Sandbrink, J.M. 1992. Clone bank and physical map of *Sedum album* chloroplast DNA. *Biochem. Syst. Ecol.* **20**: 243-253.

Ham, R.C.H.J. van, Mes, T.H.M., Hart, H. 't & Sandbrink, J.M. 1994. Molecular evolution of noncoding regions of the chloroplast genome in the Crassulaceae and related species. *Curr. Genet.* **25**: 558-566.

Hart, H. 't. 1978. Biosystematic studies in the *Acre*-group and the series *Rupestria* Berger of the genus *Sedum* L. (Crassulaceae). Ph.D. Thesis, Utrecht University.

Hart, H. 't. 1982. The white-flowered *Sedum* species. 1. Principles of a phylogenetic classification of the Sedoideae (Crassulaceae) and the position of the white-flowered *Sedum* species. *Proc. Kon. Ned. Akad. Wet., Ser. C* **85**: 663-675.

Hart, H. 't. 1985. Chromosome numbers in *Sedum* (Crassulaceae) from Greece. *Wildenowia* **15**: 115-135.

Hart, H. 't. 1991. Evolution and classification of the European *Sedum* species (Crassulaceae). *Flora Medit.* **1**: 31-61.

Hart, H. 't. 1992. The archetype of the Crassulaceae. *Acta Bot. Neerl.* **41**: 107-108.

Hart, H. 't & Koek-Noorman, J. 1989. The origin of the woody Sedoideae (Crassulaceae). *Taxon* **38**: 535-544.

Haskins, M.L. & Hayden, W.J. 1987. Anatomy and affinities of *Penthorum*. *Amer. J. Bot.* **74**: 164-177.

Hiratsuka, J., Shimada, H., Whittier, R., Ishibashi, T., Sakamoto, M., Mori, M., Kondo, C., Honji, Y., Sun, C.-R., Meng, B.-Y., Li, Y.-Q., Kanno, A., Nishizawa, Y., Hirai, A., Shinozaki, K. & Sugiura, M. 1989. The complete sequence of the rice (*Oryza sativa*) chloroplast genome: intermolecular recombination between distinct tRNA genes account for major plastid DNA inversion during the evolution of the cereals. *Mol. Gen. Genet.* **217**: 185-194.

Jansen, R.K. & Palmer, J.D. 1987. Chloroplast DNA from lettuce and *Barnadesia* (Asteraceae): structure, gene localization, and characterization of a large inversion. *Curr. Genet.* **11**: 553-564.

Jansen, R.K. & Palmer, J.D. 1988. Phylogenetic implications of chloroplast DNA restriction-site variation in the Mutisieae (Asteraceae). *Amer. J. Bot.* **75**: 753-766.

Jensen, L.C.W. 1968. Primary stem vascular patterns in three subfamilies of the Crassulaceae. *Amer. J. Bot.* **55**: 553-563.

Mauritzon, J. 1933. Studien über die Embryologie der Familien Crassulaceae und Saxifragaceae. Hakand Ohlsons, Lund.

Ohba, H. 1978. Generic and infrageneric classification of the Old World Sedoideae (Crassulaceae). *J. Fac. Sci., Univ. Tokyo, Sect. 3.* **12**: 139-198.

Ohyama, K., Fukuzawa, H., Kohchi, T., Shirai, H., Sano, T., Sano, S., Umesono, K., Shiki, Y., Takeuchi, M., Chang, Z., Aota, S., Inokuchi, H. & Ozeki, H. 1986. Chloroplast gene organization deduced from complete sequence of liverwort *Marchantia polymorpha* chloroplast DNA. *Nature* **322**: 572-574.

Olmstead, R.G. & J.D. Palmer. 1994. Chloroplast DNA systematics: a review of methods and data analysis. *Amer. J. Bot.* **81**: 1205-1224.

Palmer, J.D. 1991. Plastid chromosomes: structure and evolution. In Bogorad, L. & Vasil, I.K. (eds.), The molecular biology of plastids. Cell culture and somatic cell genetics of plants, **7A**, 5-53. Academic Press, New York.

Palmer, J.D., Jansen, R.K., Michaels, H.J., Chase, M.W. & Manhart, J.R. 1988. Chloroplast DNA variation and plant phylogeny. *Ann. Missouri Bot. Gard.* **75**: 1180-1206.

Pilon-Smits, L. 1992. Variation and evolution of crassulacean acid metabolism in *Sedum* and *Aeonium* (Crassulaceae). Ph.D. Thesis, Utrecht University.

Quimby, M.W. 1939. The floral morphology of the Crassulaceae. Ph.D. Thesis, Mississippi.

Reith, M. & Munholland, J. 1993. A high-resolution gene map of the chloroplast genome of the red alga *Porphyra purpurea*. *Plant Cell* **5**: 465-475.

Ris, H. & Plaut, W. 1962. The ultrastructure of DNA-containing areas in the chloroplast of *Chlamydomonas*. *J. Cell Biol.* **13**: 383-391.

Shinozaki, K., Ohme, M., Tanaka, M., Wakasugi, T., Hayashida, N., Matsubayashi, T., Zaita, M., Chunwongse, J., Obokata, J., Yamaguchi-Shinozaki, K., Ohto, C., Torozawa, K., Meng, B.Y., Sugita, M., Deno, H., Kamogashira, T., Yamada, K., Kusuda, J., Takaiwa, F., Kato, A., Tohdoh, N., Shimada, H. & Sugiura, M. 1986. The complete nucleotide sequence of the tobacco chloroplast genome: its gene organization and expression. *EMBO J.* **5**: 2043-2049.

Soltis, P.S., Soltis, D.E. & Doyle, J.J. 1992. Molecular systematics of plants. Chapman and Hall, New York.

Soltis, D.E., Morgan, D.R., Grable, A., Soltis, P.S. & Kuzoff, R. 1993. Molecular systematics of Saxifragaceae sensu stricto. *Amer. J. Bot.* **80**: 1056-1081.

Sugiura, M. 1992. The chloroplast genome. *Plant Mol. Biol.* **19**: 149-168.

Swofford, D.L. 1993. PAUP: Phylogenetic Analysis Using Parsimony, Version 3.1.1. Computer program distributed by the Illinois Natural History Survey. Champaign, Illinois.

Swofford, D.L. & Olsen, G.J. 1990. Phylogeny reconstruction. In Hillis, D.M. & Moritz, C. (eds.), Molecular Systematics, 411-501. Sinauer Associates, Massachusetts.

Taberlet, P., Gielly, L., Pautou, G. & Bouvet, J. 1991. Universal primers for amplification of three noncoding regions of chloroplast DNA. *Plant Mol. Biol.* **17**: 1105-1109.

Takhtajan, A. 1980. Outline of the classification of flowering plants (Magnoliophyta). *Bot. Rev.* **46**: 225-359.

Thorne, R.F. 1983. Proposed new realignments in the angiosperms. *Nordic J. Bot.* **3**: 75-117.

Uhl, C.H. 1948. Cytotaxonomic studies in the subfamilies Crassuloideae, Kalanchoideae, and Cotyledonoideae of the Crassulaceae. *Amer. J. Bot.* **35**: 695-706

Uhl, C.H. 1961a. Some cytotaxonomic problems in the Crassulaceae. *Evolution* **15**: 375-383.

Uhl, C.H. 1961b. The chromosomes of the Sempervivoideae (Crassulaceae). *Amer. J. Bot.* **48**: 114-123.

Uhl, C.H. 1963. Chromosomes and phylogeny of the Crassulaceae. *Cactus Succ. J. (US)* **35**: 80-84.

Uhl, C.H. 1992. Polyploidy, dysploidy, and chromosome pairing in *Echeveria* (Crassulaceae) and its hybrids. *Amer. J. Bot.* **79**: 556-566.

Wassmer, A. 1955. Vergleichend-morphologische Untersuchungen an den Blüten der Crassulaceen. Ph.D. Thesis, Winterthur.

Wolfe, K.H. & Sharp, P.M. 1988. Identification of functional open reading frames in chloroplast genomes. *Gene* **66**: 215-222.

Wolfe, K.H., Li, W.H. & Sharp, P.M. 1987. Rates of nucleotide substitution vary greatly among plant mitochondrial, chloroplast, and nuclear DNAs. *Proc. Natl. Acad. Sci. USA* **84**: 9054-9058.

Wolfe, K.H., Sharp, P.M. & Li, W.H. 1989. Rates of synonymous substitutions in plant nuclear genes. *J. Mol. Evol.* **29**: 208-211.

Wolfe, K.H., Morden, C.W. & Palmer, J.D. 1992. Function and evolution of a minimal plastid genome from a nonphotosynthetic parasitic plant. *Proc. Natl. Acad. Sci. USA* **89**: 10648-10652.

Zurawski, G., Clegg, M.T. & Brown, A.H.D. 1984. The nature of nucleotide sequence divergence between barley and maize chloroplast DNA. *Genetics* **106**: 735-749.

CHAPTER 3

Phylogenetic and systematic implications of chloroplast and nuclear spacer sequence variation in the Macaronesian Sempervivoideae and related Sedoideae.

Ted H.M. Mes

Abstract

Evolutionary relationships of twenty-two taxa of the Macaronesian Sempervivoideae and five related Sedoideae were inferred from sequence variation of the chloroplast DNA trnL(UAA) - trnF (GAA) intergenic spacer and from the internal transcribed spacer 2 (ITS2) of the nuclear ribosomal DNA. Phylogenetic analysis using Wagner parsimony and cluster analysis of the chloroplast and nuclear sequences revealed that the herbaceous, North African *Sedum caeruleum*, *S. pubescens*, and three species of *S.* series *Monanthoidea* are basal to a monophyletic clade containing the Macaronesian Sempervivoideae. *S.* series *Monanthoidea* is the sistergroup of the Macaronesian Sempervivoideae. The latter group comprises three monophyletic lineages, i.e. *Aeonium* (including *Greenovia*), *Aichryson*, and *Monanthes*. *Monanthes* is the sistergroup of *Aeonium*. Within *Aeonium* there are two strongly supported monophyletic clades. 1) An *Aeonium*-clade comprising the small, rosulate *A. simsii* as well as the large, woody *A. holochrysum* and *A. rubrolineatum* (Canary Islands), *A. arboreum* (Morocco), *A. gorgoneum* (Cape Verde Islands), and *A. leucoblepharum* (Yemen). 2) A *Leuconium*-clade comprising two white flowered species of *A.* sect. *Leuconium* and *A. nobile* with red flowers. The terminal position of the large woody taxa in the phylogeny strongly supports the theory that the Macaronesian Sempervivoideae evolved from a herbaceous, North African *Sedum*-like ancestor, which may have resembled the species of *Sedum* series *Monanthoidea*. Migration from North Africa to Macaronesia was accompanied by polyploidy. The systematic and nomenclatural implications of the results are summarized in a conspectus of the classification of *Aeonium* and the North African Sedoideae.

Introduction

The Macaronesian Sempervivoideae comprise approximately sixty species which are mainly restricted to the Canary Islands, Madeira, and the Azores (Macaronesia). They are generally classified in four genera, i.e., *Aeonium*, *Aichryson*, *Greenovia*, and *Monanthes* (Berger, 1930; Praeger, 1932; Bramwell, 1968, 1970; Liu, 1989; Nyffeler, 1992). *Aichryson* contains about ten, often much branched, herbaceous, predominantly hapaxanth species ranging from a few cm to 40 cm. *Greenovia* contains four rosulate species with conspicuous, highly polymerous flowers. *Monanthes* comprises nine, predominantly small, rosulate, herbaceous species with greenish or brownish flowers and conspicuously large nectariferous scales. The thirty-eight species of *Aeonium* exhibit a wide spectrum of growth-forms, ranging from small rosettes and undifferentiated (sub)shrubs to rosette trees and large, woody, much branched shrubs. *Aeonium* is often regarded as an example of adaptive radiation equal in complexity to the Darwin finches of the Galápagos islands. Thirty-two species of *Aeonium* occur on the Canary Islands, two on Madeira, one on the Cape Verde Islands, one in Western Morocco and two in Eastern Africa and South-Western Arabia (Liu, 1989).

The disjunct distribution of *Aeonium*, especially of the large, woody species of *A.* sect. *Aeonium* which occur on the Canary Islands and the Cape Verde Islands, as well as in Western Morocco, Eastern Africa and Arabia, has led to strongly contradicting theories on the evolution of this genus and the Macaronesian Sempervivoideae in general. Traditionally the four genera of the Macaronesian Sempervivoideae are regarded as descendants of woody African progenitors from the Tertiary resembling *A.* sect. *Aeonium* (Berger, 1930; Lems, 1960; Lösch, 1990). As a result of the expansion of the Sahara these ancestral species are thought to have become extinct, except for some relictual (woody) species in Eastern Africa and Western Morocco. After arrival on the Canary Islands these woody species were thought to have given rise to the present diversity through island speciation and adaptive radiation. A more recent theory implies the migration of a herbaceous ancestor to the Canary Islands which subsequently gave rise to the present diversity including the woody, Macaronesian forms (Carlquist, 1974). Some of these large, shrubby *Aeonium* species eventually remigrated to the continent (Pilon-Smits, 1992). The second scenario is primarily based on evidence obtained from wood anatomy, but is strongly supported by the presence of obligate Crassulacean Acid Metabolism (CAM) activity in the woody species. The second scenario agrees with Carlquists theory according to which the largest and woody species of otherwise predominantly herbaceous groups on oceanic islands have evolved through an increase of woodiness.

On the basis of cytological data, Uhl (1961) regarded certain North African *Sedum* species with polymerous flowers (*Sedum gattefossei* and *S. surculosum*) as related to the Macaronesian Sempervivoideae. In a chloroplast DNA restriction site analysis of Eurasian Crassulaceae, van Ham (1994) found that two North African *Sedum* species with polymerous flowers, i.e., *S. pubescens* and *S. caeruleum*, are indeed basal to the Macaronesian Sempervivoideae. More recently, phylogenetic analysis of trnL - trnF chloroplast DNA spacer sequence variation indicated that the Moroccan species of *S.* series *Monanthoidea* should also be included in a phylogenetic analysis of the Macaronesian Sempervivoideae (Mes & 't Hart, 1994).

To discriminate between the two opposing scenarios of evolution of the Macaronesian Sempervivoideae we used the chloroplast trnL (UAA) - trnF (GAA) intergenic spacer as well as the nuclear Internal Transcribed Spacer 2 (ITS2) of the Macaronesian Sempevivoideae for phylogenetic reconstruction (Baldwin, 1992; 1993). In addition to fifteen species of *Aeonium*, one species of *Greenovia* and three species each of *Aichryson* and *Monanthes*, we also included five North African *Sedum* species with polymerous flowers in this study. *Sedum sediforme* and *Sempervivum ciliosum* were used as outgroups in the chloroplast survey (van Ham, 1994).

Materials and methods

The twenty-nine species studied were all collected in nature and, except for *Sedum surculosum*, for further studies cultivated in the greenhouse of the Botanic Garden of Utrecht. DNA of *S. surculosum* was isolated from a 4 year old herbarium specimen (Herb. Rabat). The origins of the investigated plants are presented in Table 1.

Total DNA was isolated using the procedure of Guillemaut and Maréchal-Drouard (1992). 1 to 5 ng of total DNA was used for amplification. PCR conditions were 5 min denaturation (92 °C), followed by 27 cycles of 5 sec denaturation (92 °C), 30 sec annealing (55 °C), and 1 sec extension (72 °C). After cycling, extension of amplified fragments proceeded for 10 minutes at 72 °C. Direct DNA sequencing in both directions was performed with the Sanger method (1977) using the Taq Dye Terminator Cycle Sequencing Kit (ABI) and the 373 Automated DNA sequencer. Borders of the ITS2 sequences were determined by comparison with ITS, 5.8S and 25S sequences from *Daucus carota* and *Vicia faba* (Yokota & al., 1989). Primers for the trnL - trnF spacer and the ITS2 were from Taberlet & al. (1991) and White & al (1990), respectively.

The sequences were aligned with PCGENE, release 6.70, though the final alignments were edited manually. Although the ITS2 sequence of *Sempervivum ciliosum* is not used in phylogenetic analyses, this sequence is included in both alignments. Cladograms were constructed with PAUP 3.1.1. (Swofford, 1993) using unweighted Wagner parsimony. In all analyses with eleven terminal taxa, the branch and bound option of PAUP was used. When more than eleven terminal taxa were analysed, the heuristic search options nearest neighbor interchange (NNI), branch-swapping, steepest descent, and mulpars were used to search for shortest trees, and the character-state optimisation option ACCTRAN was used. Bootstrap analyses (Swofford & al., 1990; 500 replicates, 10 random addition replicates per bootstrap replicate) and decay analyses (Donoghue & al., 1992; Soltis & al., 1993) were used to assess support for monophyletic groups. Dendrograms based on Jukes-Cantor distances were generated using the Neighbor-Joining method (NJ; Saitou and Nei, 1987) as implemented in the Molecular Evolutionary Genetics Analysis package version 1.0 (MEGA; Kumar & al., 1993) with deletion of all positions containing gaps.

Table 1. Origin and accession number of the investigated plants.

Aeonium		
A. arboreum	Morocco; exact provenance unknown	HRT-31850
A. castello-paivae	Canary Islands; Gomera	HRT-30484
A. glandulosum	Madeira	HRT-30464
A. glutinosum	Madeira	HRT-30452
A. gorgoneum	Cape Verde Islands; Santo Antao	HRT-31889
A. haworthii	Canary Islands; Tenerife	HRT-31487
A. holochrysum	Canary Islands; Tenerife	HRT-30467
A. leucoblepharum	Yemen; above Sumara Pass	HRT-30669
A. nobile	Canary Islands; La Palma	HRT-32366
A. rubrolineatum	Canary Islands; Gomera	HRT-31855
A. sedifolium	Canary Islands; Tenerife	HRT-31908
A. simsii	Canary Islands; Gran Canaria	HRT-30440
A. spathulatum	Canary Islands; Tenerife	HRT-31917
A. tabuliforme	Canary Islands; Tenerife	HRT-31918
A. viscatum	Canary Islands; Gomera	HRT-30481
Aichryson		
A. dichotomum	Canary Islands; Tenerife	HRT-31938
A. punctata	Madeira	HRT-30462
A. tortuosum	Canary Islands; Lanzarote	HRT-30466
Greenovia		
G. diplocycla	Canary Islands; Tenerife	HRT-30443
Monanthes		
M. brachycaulos	Canary Islands; Gran Canaria	HRT-32180
M. Laxiflora	Canary Islands; Gomera	HRT-30486
M. polyphylla	Canary Islands; Tenerife	HRT-32187
Sedum		
S. caeruleum	Tunisia; Cap Serrat	HRT-31621
S. jaccardianum	Morocco; 34 km S of Azrou	HRT-32219
S. modestum	Morocco; Taza-Jebel Tazzeka	HRT-32226
S. pubestum	Tunisia; Jebel Ichkeul	HRT- 31627
S. surculosum	Morocco; NE of Oukaimeden	Herb. Rabat

Results

Chloroplast DNA spacer - The length of the trnL (UAA) - trnF (GAA) spacer varies among the 29 species studied from 181 bp in *Sedum surculosum* to 268 bp in *S. pubescens*. The spacer of *S. surculosum* and *S. jaccardianum* is considerably shorter than the spacer of the other taxa, which is primarily due to a 70 bp deletion (position 32 to 101). Length variation among the other taxa is rather limited and ranges from 249 bp in *Sedum modestum* and *Monanthes polyphylla* to 268 bp in *Sedum pubescens*. Spacer-length-variation in *Aeonium* is minimal (256 or 257 bp).

In the aligned chloroplast DNA trnL(UAA) - trnF(GAA) spacer sequences 19 insertions and/or deletions (indels) are present which vary in length from 1 bp to 70 bp. Only three indels are homoplasious (109, 132-138, 224), whereas ten indels are autapomorphies and six are synapomorphies. A noteworthy indel is a putative 1 bp deletion (224) shared by the five woody species of *Aeonium* sect. *Aeonium* and the rosulate, monopodial *A. simsii.* Excluding alignment gaps, 81 variable positions (base substitutions) are present. Of these, 13 are outgroup variable positions, 37 are autapomorphies, 12 are ingroup/outgroup variable positions, and 19 are synapomorphies. In the sixteen species of *Aeonium* (including *Greenovia diplocycla*), only four variable sites were observed of which only one is cladistically informative, i.e. a synapomorphy at position 205 for *Aeonium sedifolium* and *Greenovia diplocycla.* Given the relative lack of intrageneric sequence variation of the chloroplast spacer in the Macaronesian genera, only a single representant of each of the four genera, together with all *Sedum* species were used to perform cluster analyses and cladistic analyses.

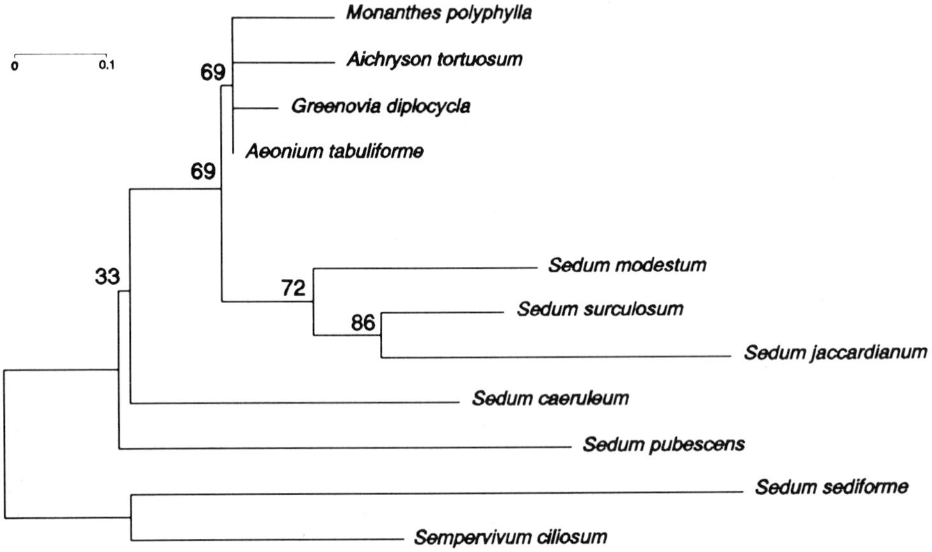

Figure 1. Neighbor-Joining dendrogram constructed from Jukes-Cantor distances for the trnL (UAA) - trnF (GAA) intergenic chloroplast DNA spacer of Macaronesian Sempervivoideae and five North African *Sedum* species. *Sedum sediforme* and *Sempervivum ciliosum* were used as outgroups. The numbers above the branches are bootstrap values (500 replicates).

The Neighbor-Joining tree of the chloroplast DNA sequences of 11 species of Sempervivoideae and Sedoideae (Fig. 1) reveals that *Sedum sediforme* and *Sempervivum ciliosum* are genetically most distantly related to all other taxa and consequently these species are indeed appropriate outgroups for the Macaronesian Sempervivoideae and related North African Sedoideae as has been indicated by van Ham (1994). In the Neighbor-Joining dendrogram the North African *Sedum caeruleum* and *S. pubescens* are basal to the Macaronesian Sempervivoideae as well as to *S.* series *Monanthoidea,* but the branching order of the two *Sedum* species is poorly resolved. *S.* series *Monanthoidea*, which contains *S. modestum*, *S. jaccardianum* and *S. surculosum*, is the sistergroup of the Macaronesian Sempervivoideae. Within *S.* series *Monanthoidea*, *S. modestum* is most frequently basal to *S. surculosum* and *S. jaccardianum*. Wagner parsimony analysis of chloroplast DNA sequences of these eleven species (Fig. 2) found 1 shortest tree (length 79 steps; consistency 0.91). Using *Sempervivum ciliosum* and *Sedum sediforme* as outgroups, again *S. pubescens* and *S. caeruleum* are basal to *S.* series *Monanthoidea* and the Macaronesian Sempervivoideae. The four genera of the Macaronesian Sempervivoideae together with *Sedum* series *Monanthoidea* form a monophyletic clade but the relationships among the five taxa are unresolved. Bootstrap analysis indicated that the ingroup as a whole and *S.* series *Monanthoidea* are the strongest supported clades.

Internal Transcribed Spacer 2 (ITS2) - The length of the ITS2 region in the Macaronesian Sempervivoideae and related Sedoideae ranges from 203 bp in *Sedum surculosum* to 214 bp in *Monanthes laxiflora*. Intrageneric variation in spacer length is 212-213 bp in *Aeonium*, 210-211 bp in *Aichryson*, 212-214 bp in *Monanthes*, and 203-209 bp in *Sedum* series *Monanthoidea*. The ITS2 spacer length of *S. caeruleum* and *S. pubescens* is 213 and 211 bp, respectively.

In the aligned ITS2 sequences 31 non-overlapping indels are present of which only four indels are longer than 1 bp (28-32, 183-186, 183-188, and 210-211). In the ITS2 sequences most variation is accounted for by base substitutions. In the ITS2 sequences of the taxa examined, 127 variable positions are present (gaps and outgroup variable positions excluded). The Macaronesian Sempervivoideae contain a large number of variable sites, i.e. 26 positions in the 16 species of *Aeonium* (including *Greenovia*), 15 positions in the three species of *Aichryson*, and 15 positions in the three taxa of *Monanthes*. The five *Sedum* species comprise 80 variable positions, most of which were found in *S. pubescens* and *S. caeruleum*. *S. modestum*, *S. jaccardianum*, and *S. surculosum* comprise only 18 variable sites.

Phylogenetic analysis of chloroplast DNA restriction site variation indicates that *Sedum pubescens*, *S. caeruleum*, and the Macaronesian Sempervivoideae constitute a strongly supported, monophyletic clade within the Crassulaceae (van Ham, 1994). Our analysis of the trnL - trnF chloroplast DNA spacer indicates that the genetic divergence between these two *Sedum* species and the ingroup taxa exceeds all other combinations. We therefore used *Sedum caeruleum* and *S. pubescens* as outgroups in phylogenetic analyses of the ITS2 sequence data. The strict consensus tree of the ITS2 data (consistency 0.78) of 67 shortest trees with a length of 239 which resulted from a heuristic unweighted Wagner analysis, treating gaps as missing data with presence/absence coding of indels (Wojciechowski & al., 1993) is shown in Figure 3. The topology of this cladogram is very similar to the Neighbor-Joining tree (not shown). The cladistic analysis indicates that 1) *S.* series *Monanthoidea* is monophyletic and that this series is the sistergroup of the Macaronesian Sempervivoideae, 2) the Macaronesian Sempervivoideae are monophyletic and comprise three monophyletic lineages, i.e. *Aeonium* (including *Greenovia*), *Aichryson*, and *Monanthes*, 3) the five species of *Aeonium* sect.

Aeonium and *A. simsii* form a monophyletic, terminal clade within *Aeonium* and 4) *A. castello-paivae* and *A. haworthii* (*A.* sect. *Leuconium*) and *A. nobile* (*A.* sect. *Megalonium*) also constitute a monophyletic clade within *Aeonium.*

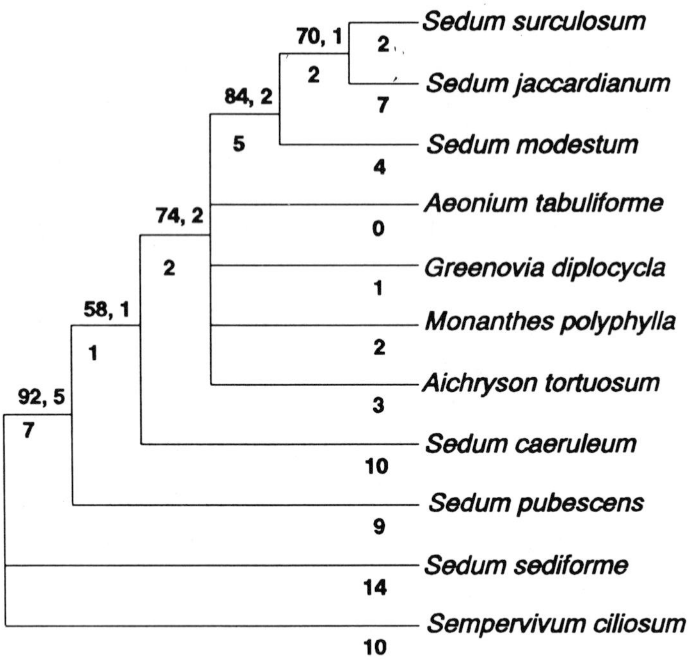

Figure 2. Single shortest Wagner cladogram (71 steps, consistency 0.91) based on substitutions of the chloroplast trnL (UAA) - trnF (GAA) intergenic chloroplast DNA spacer of four Macaronesian Sempervivoideae and five North African *Sedum* species. *Sedum sediforme* and *Sempervivum ciliosum* were used as outgroups. Bootstrap (500 replicates) and decay values are shown above branches, branch lengths below branches.

Discussion

Phylogenetic analysis of the chloroplast and nuclear encoded sequences results in highly congruent phylogenetic trees, though these particular DNA spacers have very dissimilar mechanisms of evolution. The chloroplast DNA sequences indicate that *Sedum caeruleum* and *S. pubescens* are basal to *S.* series *Monanthoidea* and the Macaronesian Sempervivoideae. The monophyletic *S.* series *Monanthoidea* is the sistergroup of the Macaronesian Sempervivoideae (Fig. 1). *Aeonium, Aichryson, Greenovia,* and *Monanthes* also form a monophyletic group but the relationships among these four genera are not resolved.

Genetic diversity based on ITS2 sequences is highest between *Sedum caeruleum* and *S. pubescens* on one hand and the other species on the other. This supports their use as outgroups. Phylogenetic analysis using Wagner parsimony (Fig. 2) supports the sister-group relationship of *S.* series *Monanthoidea* relative to the Macaronesian Sempervi-

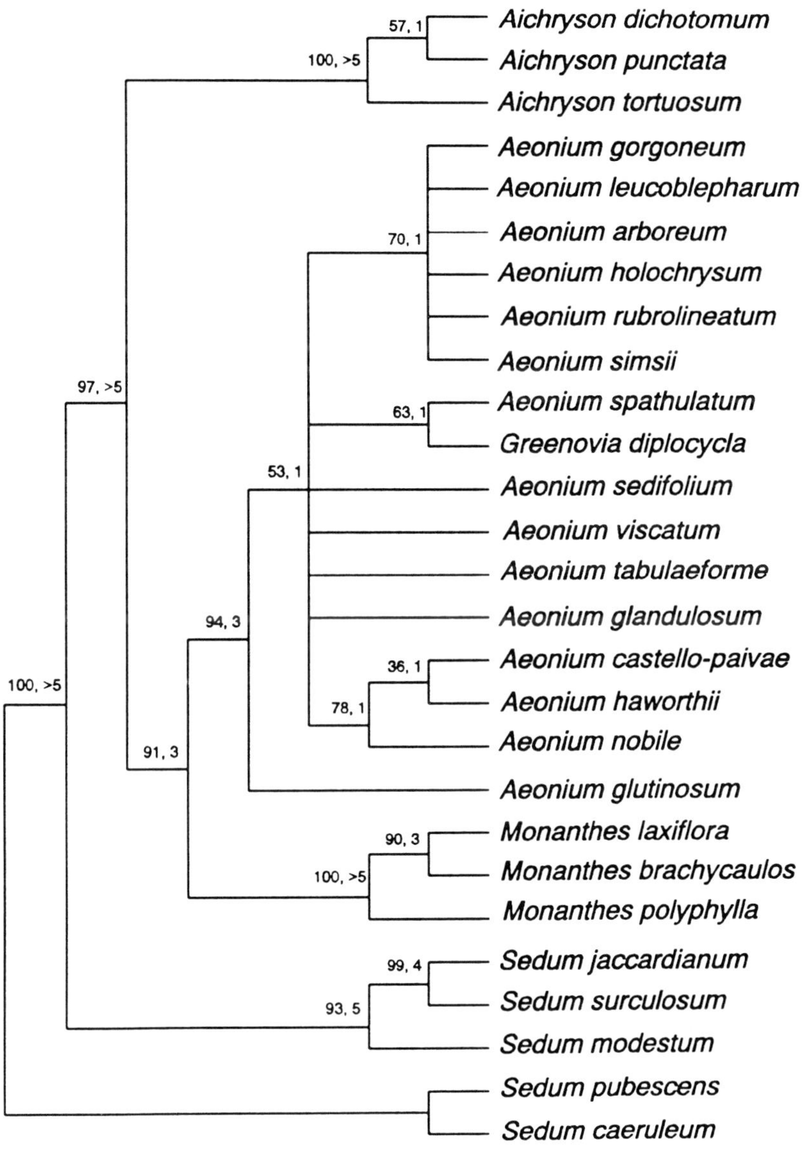

Figure 3. Strict consensus of 67 most parsimonious unweighted Wagner cladograms (239 steps, consistency 0.78) based on substitutions of the Internal Transcribed Spacers 2 (ITS2) of twenty-two Macaronesian Sempervivoideae and five North African *Sedum* species. Bootstrap values (500 replicates) and decay values are shown above the branches.

voideae and indicates the monophyly of each of the genera *Aichryson*, *Aeonium* (*Greenovia* included), and *Monanthes*. The relationships among the genera are much more resolved when compared to the chloroplast DNA based phylogeny: *Monanthes* is the sistergroup of *Aeonium* (including *Greenovia*) and *Aichryson* is basal to *Aeonium* and *Monanthes*.

Both chloroplast and nuclear sequences indicate that the Macaronesian Sempervivoideae are monophyletic and that the North African *Sedum caeruleum*, *S. pubescens*, and *S.* series *Monanthoidea* are basal to this group (Fig. 3). It is therefore most likely that the Macaronesian Sempervivoideae evolved from a North African *Sedum*-like ancestor. Because *Aeonium* is the only genus of the Macaronesian Sempervivoideae which is distinctly woody, and because its sistergroup (*Monanthes*) as well as the Macaronesian (*Aichryson*) and North African groups (*Sedum*) basal to *Aeonium* are predominantly herbaceous, the ancestor of the Macaronesian Sempervivoideae was almost certainly also herbaceous. As *Sedum* series *Monanthoidea* is the sistergroup of the Macaronesian Sempervivoideae it is tempting to assume that the ancestor of the Macaronesian Sempervivoideae resembled the species of *S.* series *Monanthoidea*.

The ancestor of the Macaronesian Sempervivoideae was probably diploid because the *Sedum* species at the base of the phylogeny are diploid with basic numbers ranging from x=6 in *Sedum pubescens* and x=8 in *S. jaccardianum*, *S. modestum*, and *S. surculosum* to x=12-13 in *S. caeruleum* (Hébert, 1967; Grant, 1971; Uhl, unpublished data; 't Hart, unpublished data). The only exception is *S. surculosum* ssp. *surculosum* in which the dysploid polyploid basic number x=15 has been found (Galland, 1988; Mes & 't Hart, 1994). Since all Macaronesian Sempervivoideae are polyploid with secondary basic numbers ranging from x=15, 16, and 17 in *Aichryson* to x=18 in *Aeonium*, *Greenovia*, and *Monanthes*, the migration from the African mainland to Macaronesia was associated with polyploidy. The increase of chromosome numbers must have occurred either on the continent in an extinct *Sedum*-like ancestor or on the Macaronesian islands. Migration to oceanic islands associated with polyploidization followed by adaptive radiation and island speciation without further change of chromosome numbers has been reported for several taxa of the flora of the Hawaiian Islands and the Juan Fernandez Islands (Helenurm & Ganders, 1985; Sanders & al., 1983; Stuessy & al., 1992). A similar pattern occurs in the Macaronesian Sempervivoideae and the related North African *Sedum* taxa, though secondary tetraploid chromosome numbers have been reported for some species, e.g. 2n=60 for *Aichryson villosum*, 2n=64 for *A. pachycaulon*, and 2n=72 for *Aeonium arboreum*, *A. balsamiferum*, *A. haworthii*, *A. holochrysum*, *A. leucoblepharum*, *A. simsii*, *A. undulatum*, *A. urbicum*, and *A. valverdense* as well as for *Greenovia diplocycla*, *Monanthes anagensis*, and *M. polyphylla*.

The Macaronesian Sempervivoideae have developed a wide range of different growth-forms through adaptive radiation after their arrival in Macaronesia. The phylogeny based on nuclear DNA sequences clearly indicates that the large, branched woody forms of *Aeonium* evolved in Macaronesia. These results fully support Carlquists (1962, 1974) theory on the origin of woody growth-forms on oceanic islands. The presence of a few woody taxa of *A.* sect. *Aeonium* on the African continent, e.g. *A. arboreum* (*A. korneliuslemsii*) in Morocco and *A. leucoblepharum* and *A. stuessyi* in

Eastern Africa and Yemen must be the result of relatively recent remigration of members of this section through long distance dispersal.

The relationships among the 16 species of *Aeonium* (including *Greenovia*) studied here are incompletely resolved. Only two clades are relatively well supported, i.e. a *Leuconium*-clade comprising *A. nobile* together with *A. castello-paivae* and *A. haworthii* of *A.* sect. *Leuconium* and an *Aeonium*-clade comprising *A. simsii* together with *A. arboreum*, *A. gorgonium*, *A. holochrysum*, *A. leucoblepharum*, and *A. rubrolineatum* of *A.* sect. *Aeonium*. These species of the *Aeonium*-clade are characterised by a unique indel (1 bp) in both the chloroplast spacer and nuclear spacer. Within the *Leuconium*- and *Aeonium*-clades, sequence variation of the ITS2 is very low. There are no variable sites among the three species of the *Leuconium*-clade and only two variable sites among the six species of the *Aeonium*-clade (gaps excluded). The remaining seven species of *Aeonium* in this study, on the other hand, comprise 18 variable positions. These figures indicate that whereas maximum genetic divergence of the ITS2 amounts to 5.5% in this sample of *Aeonium*, genetic divergence is essentially absent in the *Leuconium*- and *Aeonium*-clades (sequence divergence $\leq$ 1.0%). The lack of genetic variation within these two distinct clades may indicate extremely rapid morphological adaptation, e.g. the *Aeonium*-clade comprises the herbaceous, rosulate *A. simsii* as well as the large subshrubs of *A.* sect. *Aeonium*.

Taxonomy

The clades distinguished in the phylogenetic analysis of the chloroplast and nuclear encoded sequences of the Macaronesian Sempervivoideae and the related North African Sedoideae do not fully agree with the present classifications of these taxa. Especially the infrageneric classification of *Aeonium*, the position of *Greenovia*, and the delimitation of infrageneric groups in North African *Sedum* need to be revised.

The sections of *Aeonium* are primarily based on growth-form and flower morphology. The four most recent monographs of *Aeonium* roughly distinguish four major groups (Berger, 1930; Praeger, 1932; Lems, 1960; Liu, 1989): 1) The hapaxanth, rosulate species of *A.* sect. *Canariensia* (*A.* sect. *Patinaria*). 2) The large, woody, shrubby taxa with yellow flowers of *A.* sect. *Aeonium* (*A.* sect. *Holochrysa*). Liu (1989) classified the shrubby species of Madeira, the Cape Verde Islands and Eastern Africa in *A.* sect. *Pittonium*. 3) The large or medium sized, woody species with large, often urceolate, white flowers of *A.* sect. *Leuconium*. 4) The herbaceous or small, twiggy shrublets with yellow flowers (only *A. goochiae* has small whitish flowers). Praeger (1932) and Lems (1960) classified these species all in *A.* sect *Goochiae*. Berger (1930) and Liu (1989), on the other hand, placed these species in three and two sections, respectively. In addition to these four major groups all authors distinguish a monotypic *A.* sect. *Megalonium* containing the red flowered *A. nobile*. Our results strongly support the inclusion of *A. simsii* and *A. nobile* in *A.* sect. *Aeonium* and *A.* sect. *Leuconium*, respectively. Furthermore, *A. glandulosum* is excluded from *A.* sect. *Canariensia* and *A. glutinosum* is neither related to the species of *A.* sect. *Aeonium* (Liu, 1989) nor to *A.* sect. *Canariensia*

(Praeger, 1932).

Greenovia comprises four rosulate species with yellow 20- to 32-merous flowers. The plants are hapaxanth or produce stolons from the lower rosette leaves. *Greenovia* is generally regarded as closely related to *Aeonium*, especially to *A.* sect. *Canariensia* (Berger, 1930; Praeger, 1932; Liu, 1989). However, the highly polymerous flowers of *Greenovia*, the absence of nectariferous scales and a different placentation (Berger, 1930; Liu 1989) have been considered important enough for generic status. Our results, however, unequivocally support the inclusion of *Greenovia* in *Aeonium*.

The North African *Sedum* species with polymerous flowers (excluding species of the *Sedum rupestre* group), have been classified in various numbers of infrageneric taxa of different ranks (Battandier & Trabut, 1899; Berger, 1930; Fröderström, 1932; Maire, 1977). The classification presented here is in accordance with a classification of the European *Sedum* species ('t Hart, 1991).

Conspectus

(species marked with an asterisk have been used in the present phylogenetic analysis)

Aeonium Webb & Berth. Lectotype, *Aeonium arboreum* (L.) Webb & Berth. (Liu in Syst. *Aeonium*, 35. 1989).

Aeonium sect. *Aeonium*
Synonyms: *Sempervivum* sect. *Aeonium* [subsect.] *Holochrysa* Christ in Bot. Jahrb. Syst. 9, 118. 1888; *Aeonium* sect. *Euaeonium* A. Berger in Engler & Prantl, Nat. Pflanzenfam. ed 2., 18a, 431. 1930 (nom. inval.); *Aeonium* sect. *Holochrysa* (Christ) Praeger in Account of *Sempervivum* 152. 1932. Type, *Aeonium holochrysum* Webb & Berth.; *Aeonium* sect. *Chrysocome* series *Simsii* Liu in Syst. Aeonium, 46. 1989. Type, *Aeonium simsii* (Sweet) Stearn.

Species included: **A. arboreum* (L.) Webb & Berth., *A. balsamiferum* Webb & Berth., **A. gorgoneum* J.A. Schmidt, *A. korneliuslemsii* Liu, **A. leucoblepharum* Webb ex A. Rich., **A. holochrysum* Webb & Berth., **A. rubrolineatum* Sventenius, *A. simsii* (Sweet) Stearn, *A. stuessyi* Liu, *A. undulatum* Webb & Berth.

Aeonium sect. *Goochiae* (Christ) Praeger in Account of *Sempervivum*, 191. 1932.
Basionym: *Sempervivum* sect. *Goochiae* Christ in Bot. Jahrb. Syst. 9, 117. 1888. Type, *Aeonium lindleyi* Webb & Berth.
Synonym: *Aeonium* sect. *Anodontonium* Berger in Engler & Prantl, Nat. Pflanzenfam. ed. 2, 18a, 426. 1930. Lectotype, *Aeonium lindleyi* Webb & Berth. (Liu in Syst. *Aeonium*, 39. 1989).

Species included: *A. lindleyi* Webb & Berth., *A. goochiae* Webb & Berth., **A. viscatum* Webb ex Bolle.

Aeonium sect. *Petrothamnium* (Webb) Liu in Syst. *Aeonium*, 39. 1989.

Basionym: *Greenovia* sect. *Petrothamnium* Webb in Christ in Bot. Jahrb. Syst. 9, 115. 1888. Type, *Greenovia sedifolia* (Webb ex Bolle) Christ.

Species included: *A. saundersii* Bolle, **A. sedifolium* (Webb ex Bolle) Pit. & Proust.

Aeonium sect. *Canariensia* (Christ) Praeger in Account of *Sempervivum*, 129. 1932.
Basionym: *Sempervivum* sect. *Aeonium* [subsect.] Canariensia Christ in Bot. Jahrb. Syst. 9, 118. 1888. Type, *Sempervivum canariense* L.
Synonym: *Aeonium* sect. *Patinaria* series *Patinaria* Liu in Syst. *Aeonium*, 54. 1989. Type, *Aeonium canariense* (L.) Webb & Berth.

Species included: *A. canariense* (L.) Webb & Berth., *A. cuneatum* Webb & Berth., *A. palmense* Webb ex Christ, *A. subplanum* Praeger, **A. tabuliforme* (Haw.) Webb & Berth., *A. virgineum* Webb ex Christ.

Aeonium sect. *Greenovia* (Webb & Berth.) T. Mes, comb. et stat. nov.
Basionym: *Greenovia* Webb & Berth. in Phyt. Can. 1, 198. 1840. Lectotype (designated here), *Greenovia aurea* (Chr. Smith ex Hornem.) Webb & Berth.
Synonym: *Sempervivum* sect. *Greenovia* (Webb & Berth.) Schönland in Engler & Prantl, Nat. Pflanzenfam. 3, 2a, 32. 1891.

Species included: *Aeonium aizoon* (Bolle) T. Mes, comb. nov. (Basionym, *Greenovia aizoon* Bolle in Bonplandia 7, 242. 1859), *Aeonium aureum* (Chr. Smith ex Hornem.) T. Mes, comb. nov. (Basionym, *Sempervivum aureum* Chr. Smith ex Hornem. in Enum. Pl. Hort. Hafn., suppl., 61. 1819), *Aeonium diplocyclum* (Webb ex Bolle) T. Mes, comb. nov. (Basionym, *Greenovia diplocycla* Webb ex Bolle in Bonplandia 7, 242. 1859), *Aeonium dodrentale* (Willd.) T. Mes, comb. nov. (Basionym, *Sempervivum dodrentale* Willd. in Enum. Pl. Hort. Reg. Berol. 1, 508. 1809).

Aeonium sect. *Chrysocome* Webb in Christ in Bot. Jahrb. Syst. 9, 109. 1888. Lectotype (designated here), *A. cruentum* Webb & Berth. (= *A. spathulatum*) [Liu in Syst. *Aeonium*, 47, 1989, designated *A. spathulatum* as lectotype, but Christ did not include the latter in section *Chrysocome*]
Synonyms: *Aeonium* sect. *Auonium* Berger in Engler & Prantl, Nat. Pflanzenfam. ed. 2, 18a, 427. 1930. Lectotype, *A. spathulatum* (Hornem.) Praeger (Liu, 1989); *Aeonium* sect. *Trichonium* Berger in Engler & Prantl, Nat. Pflanzenfam. ed. 2, 18a, 428. 1930. Type, *A. smithii* (Sims) Webb & Berth.

Species included: *A. smithii* (Sims) Webb & Berth., **A. spathulatum* (Hornem.) Praeger.

Aeonium sect. *Leuconium* Berger in Engler & Prantl, Nat. Pflanzenfam. ed. 2, 18a, 429. 1930. Lectotype, *A. ciliatum* (Willd.) Webb & Berth. (Liu in Syst. *Aeonium*, 78. 1989).
Synonyms: *Sempervivum* sect. *Aeonium* [subsect.] *Urbica* Christ in Bot. Jahrb. Syst. 9, 118. 1888. Type, *Aeonium urbicum* (Chr. Smith ex Hornem.) Webb & Berth.; *Aeonium* sect. *Urbica* (Christ) Praeger in Account of *Sempervivum* 166. 1932;

Aeonium sect. *Megalonium* Berger in Engler & Prantl, Nat. Pflanzenfam. ed. 2, 18a, 428. 1930. Type *Aeonium nobile* (Praeger) Praeger.

Species included: **A. castello-paivae* Bolle, *A. ciliatum* (Willd.) Webb & Berth., *A. decorum* Webb ex Bolle, *A. davidbramwellii* Liu, *A. gomerense* (Praeger) Praeger, **A. haworthii* Salm-Dyck ex Webb & Berth., *A. hierrense* (Murray) Pit. & Proust, *A. lancerottense* (Praeger) Praeger, **A. nobile* (Praeger) Praeger, *A. percarneum* (Murray) Pit. & Proust, *A. urbicum* (Smith ex Hornem.) Webb & Berth., *A. valverdense* (Praeger) Praeger.

Aeonium sect. *Patinaria* (Lowe) Berger in Engler & Prantl, Nat. Pflanzenfam. ed. 2, 18a, 428. 1930.

Basionym: *Sempervivum* [sect.] *Patinaria* Lowe in Man. Fl. Madeira 332. 1864. Type, *Sempervivum glandulosum* Aiton.

Species included: **A. glandulosum* (Aiton) Webb & Berth.

Aeonium sect. *Pittonium* Berger in Engler & Prantl, Nat. Pflanzenfam. ed. 2, 18a, 428. 1930. Type, *Aeonium glutinosum* (Aiton) Webb & Berth.

Species included: **A. glutinosum* (Aiton) Webb & Berth.

Sedum L.

Sedum series *Caerulea* Fröderström in Acta Horti. Gothob. 7, app. 103. 1932. Lectotype, *Sedum caeruleum* L. (ICBN, Art. 22.5, 1994; 't Hart in Fl. Mediter. 1, 55. 1991).

Synonym: *Sedum* sect. *Epeteium* subsect. *Caerulea* Maire in Fl. Afr. Nord 14, 378. 1977 (nom. nud.).

Species included: **S. caeruleum L.*

Sedum series *Monanthoidea* Batt. ex Fröderström in Acta Hot. Gotob. 7, app. 108, 1932 (see Ohba in J. Fac. Sci., Univ. Tokyo, S. 3, 12, 142. 1978). Lectotype (designated here), *Sedum surculosum* Coss.

Basionym: *Sedum* sect. *Monanthoidea* Battandier in Battandier & Trabut, Fl. Algérie 1, 328. 1889 (nom. nud.).

Synonyms: *Sedum* sect. *Monanthella* Berger in Engler & Prantl, Nat. Pflanzenfam. ed. 2, 18a, 446. 1930 (nom. inval.). *Sedum* sect. *Epeteium* subsect. *Modesta* Maire in Fl. Afr. Nord 14, 382. 1977 (nom. nud.).

Species included: **S. jaccardianum* Maire & Wilczek, **S. modestum* Ball, **S. surculosum* Coss.

Sedum series *Pubescens* T. Mes, series novo.

Synonyms: *Sedum* sect. *Epeteium* subsect. *Mauretanica* Maire in Fl. Afr. Nord 14, 384. 1977 (nom. nud.).

Differt a seribus praecedentibus plantis tantum annuis, foliis teretibus vel semiteretibus, sessilibus, oblongis vel linearibus, obtusis, floribus 6(5-7)-meris. Typus: *Sedum pubescens* Vahl.

Species included: **S. pubescens* Vahl.

Acknowledgements

I would like to thank Gea Zijlstra for comments and suggestions concerning nomenclature and (lecto-)typification.

References

Baldwin, B.G. 1992. Phylogenetic utility of the internal transcribed spacers of nuclear ribosomal DNA in plants: an example from the Compositae. *Molec. Phylogen. Evol.* **1**: 3-16.

Baldwin, B.G. 1993. Molecular phylogenetics of *Calycadenia* (Compositae) based on ITS sequences of nuclear ribosomal DNA: chromosomal and morphological evolution reexamined. *Amer. J. Bot.* **80**: 222-238.

Battandier, J.A. & Trabut, L.C. 1888-1889. Flore de l'Algerie. **1**. A. Jourdan, Algiers.

Berger, A. 1930. Crassulaceae. In Engler, A. & Prantl, K. (eds.), Die natürlichen Pflanzenfamilien, 2nd ed., **18a**, 352-483. W. Engelmann, Leipzig.

Bramwell, D. 1968. Notes on the taxonomy and nomenclature of the genus *Aichryson. Bol. Inst. Nat. Inv. Agron., Madrid* **28**: 203-213.

Bramwell, D. 1970. Generic delimitation of the *Sempervivum* group. *Nation. Cact. Succ. J.* **25**: 50-51.

Carlquist, S. 1962. A theory of paedomorphosis in dicotyledonous woods. *Phytomorphology* **12**: 30-45.

Carlquist, S. 1974. Island Biology. Columbia University Press, New York.

Donoghue, M.J., Olmstead, R.G., Smith, J.F. & Palmer, J.D. 1992. Phylogenetic relationships of Dipsacales based on rbcL sequences. *Ann. Missouri Bot. Gard.* **79**: 333-345.

Fröderström, H. 1932. The genus *Sedum* L. 3. *Acta Hort. Gothob.* **7**(app.): 1-126.

Galland, N. 1988. Recherche sur l'origine de la flore orophile du Maroc: Etude caryologique et cytogeographique. *Trav. Inst. Sci., Rabat, Sér. Bot.* **35**: 1-168.

Grant, V. 1971. Plant speciation. Columbia University Press, New York.

Guillemaut, P. & Maréchal-Drouard, L. 1992. Isolation of plant DNA: a fast, inexpensive, and reliable method. *Plant Mol. Biol. Rep.* **10**: 60-65.

Ham, R.C.H.J. van. 1994. Phylogenetic implications of chloroplast DNA variation in the Crassulaceae. Ph.D. Thesis, Utrecht University.

Ham, R.C.H.J. van, Hart, H. 't, Mes, T.H.M. & Sandbrink, J.M. 1994. Molecular evolution of noncoding regions of the chloroplast genome in the Crassulaceae and related species. *Curr. Genet.* **25**: 558-566.

Hart, H.'t. 1991. Evolution and classification of the European *Sedum* species (Crassulaceae). *Fl. Medit.* **1**: 31-61.

Hébert, L.P. 1976. Nouvelle contribution à la cytotaxonomie du genre *Sedum* L. *Bull. Soc. Neuchâtel. Sci. Nat.* **99**: 97-107.

Helenurm, K. & Ganders, F.R. 1985. Adaptive radiation and genetic differentiation in Hawaiian *Bidens. Evolution* **39**: 753-765.

Kumar, S., Tamura, K. & Nei, M. 1993. MEGA: Molecular Evolutionary Genetics Analysis, version 1.0. The Pennsylvania State University, Park.

Lems, K. 1960. Botanical notes on the Canary Islands. II. The evolution of plants forms in the islands: *Aeonium. Ecology* **41**: 1-47.

Liu, H.-Y. 1989. Systematics of *Aeonium* (Crassulaceae). National Museum of Natural Science, Taiwan.

Lösch, R. 1990. Funktionelle Voraussetzungen der adaptiven Nischenbesetzung in der Evolution der makaronesischen Semperviven. *Diss. Bot.* **146**.

Maire, R. 1977. Flore de l'Afrique du Nord **14**. Éditions Lechevalier, Paris.

Mes, T.H.M. & Hart, H. 't. 1994. *Sedum surculosum* and *S. jaccardianum* (Crassulaceae) share a unique 70 bp deletion in the chloroplast DNA trnL (UAA) - trnF (GAA) intergenic spacer. *Pl. Syst. Evol.* **193**: 213-221.

Nyffeler, R. 1992. A taxonomic revision of the genus *Monanthes* Haworth (Crassulaceae). *Bradleya* **10**: 49-82.

Pilon-Smits, E.A.H. 1992. Variation and evolution of Crassulacean Acid Metabolism in *Sedum* and *Aeonium* (Crassulaceae). Ph.D. Thesis, Utrecht University.

Praeger, R.L. 1932. An account of the *Sempervivum* group. Royal Horticultural Society, London.

Saitou, N. & Nei, M. 1987. The neighbor-joining method: a new method for reconstructing phylogenetic trees. *Mol. Biol. Evol.* **9**: 945-967.

Sanders, R.W., Stuessy, T.F. & Rodrigues, R. 1983. Chromosome numbers from the flora of the Juan Fernandez Islands. *Amer. J. Bot.* **70**: 799-810.

Sanger, F., Nicklen, S. & Coulsen, A.R. 1977. DNA-sequencing with chain-terminating inhibitors. *Proc. Natl. Acad. Sci. USA* **74**: 5463-5467.

Soltis, D.E., Morgan, D.R., Grable, A., Soltis, P.S., & Kuzoff, R. 1993. Molecular systematics of Saxifragaceae sensu stricto. *Amer. J. Bot.* **80**: 1056-1081.

Stuessy, T.F., Crawford, D.J. & Silva, M.O. 1992. Evolution of the endemic vascular flora of the Juan Fernandez Islands. In Dudley, E.C. (ed.). The unity of evolutionary biology, 337-339. Dioscorides Press, Portland.

Swofford, D.L. 1993. PAUP: Phylogenetic Analysis Using Parsimony, Version 3.1.1. Computer program distributed by the Illinois Natural History Survey. Champaign, Illinois.

Swofford, D.L. & Olsen G.J. 1990. Phylogeny construction. In Hillis, D.M. & Moritz, C. (eds), Molecular systematics, 411-501. Sinauer Associates, Massachusetts.

Taberlet, P., Gielly, L., Pautou, G. & Bouvet, J. 1991. Universal primers for amplification of three noncoding regions of chloroplast DNA. *Plant Mol. Biol.* **17**: 1105- 1109.

Uhl, C.H. 1961. The chromosomes of the Sempervivoideae (Crassulaceae). *Amer. J. Bot.* **48**: 114-123.

White, T.J., Bruns, T., Lee, S. & Taylor, J. 1990. Amplification and direct sequencing of fungal ribosomal RNA genes for phylogenetics. In Innis, M., Gelfand, D., Sinnsky, J. & White, T. (eds.), PCR protocols: a guide to methods and applications, 315-322. Academic Press, San Diego.

Wojciechowski, M.F., Sanderson, M.J., Baldwin, B.G., & Donoghue, M.J. 1993 Monophyly of aneuploid *Astragalus* (Fabaceae): evidence from nuclear ribosomal DNA internal transcribed spacer sequences. *Amer. J. Bot.* **80**: 711- 722.

Yokota, Y., Kawata, T., Iida, Y., Kato, A. & Tanifuji, S. 1989. Nucleotide sequences of the 5.8S rRNA gene and internal transcribed spacer regions in carrot and broad bean ribosomal DNA. *J. Mol. Evol.* **29**: 294-301.

CHAPTER 4

Chemotaxonomy of the Eurasian Sedoideae and Sempervivoideae

Jan F. Stevens

Abstract

Phytochemistry is a valuable source of comparative data to help solve taxonomic and evolutionary problems, mostly at lower hierarchical levels. This particular field of plant taxonomy is generally referred to as chemotaxonomy, chemosystematics, (bio)chemical systematics, or micromolecular systematics. In our phytochemical studies we examined the alkaloids, tannins, flavonoids, leaf wax alkanes, and epicuticular terpenoids of the Eurasian Sedoideae and Sempervivoideae with the aim to elucidate phylogenetic relationships within and between these taxa. The presence of alkaloids appeared to be strongly correlated with a phylogeny of the genus based on molecular characters, and was found to be limited to the so-called *Acre*-lineage, which comprises the Asian, Eurasian, and American *Sedum* species with a reticulate testa as well as the Central American Sedoideae and Echeverioideae. Alkaloids are absent from the remainder of *Sedum* and also from the Sempervivoideae. These taxa are endowed with large amounts of condensed tannins and galloyl esters. Given the fact that these tannins are widely distributed in the Hamamelididae-Dilleniidae-Rosidae complex, which comprises the Crassulaceae, their presence probably indicates the primitive condition in the family. 8-Hydroxy and 8-methoxy flavonols were detected in most taxa of *Sedum*, but no clear evolutionary information could be inferred from their distribution, which is probably due to a high degree of parallel evolution. *Sempervivum* appeared to be very uniform in its flavonoid chemistry with kaempferol as the principal flavonol. An investigation of the leaf exudates of *Aeonium* revealed the presence of a wide variety of flavonoid aglycones, which are taxonomically important at the infrageneric level. In addition, epicuticular terpenoids proved to be valuable taxonomic markers in *Sedum* and *Aeonium*. Glaucous species of *Sedum* usually contain large amounts of epicuticular triterpenes, which are virtually absent from *Sempervivum* and *Aeonium*. In the latter, triterpenes have been replaced by diterpenes.

Introduction

Secondary plant constituents were regarded as being waste products for a long time, but more recently their ecological and physiological functions have been widely acknowledged. In fact, the production and accumulation of secondary plant metabolites constitutes an important defense mechanism against herbivore pests and pathogens. The interaction between plants and butterflies inspired Ehrlich & Raven (1965) to postulate a new theory of biochemical coevolution, which became known as the 'coevolutionary arms race' theory. In short, it states that the chemical patterns of plant toxins constantly change during evolution as a consequence of variations in insect feeding pressures. Groups of defence agents may be replaced by others, which - at least in part - accounts for the discontinuous distribution of many classes of secondary metabolites. This must be kept in mind when using secondary plant products as chemical characters to solve taxonomic problems (Harborne & Turner, 1984).

Plants may utilize any of the three main biosynthetic routes in order to produce toxic defence agents. These are the alkaloid (nitrogen or amino acid), the terpenoid (mevalonate or isoprene), and the phenolic (shikimate) pathway. Although the phenolic pathway is operative in all major taxa of higher plants, polymeric or oligomeric flavanols (condensed tannins) are mainly found in the gymnosperms and woody angiosperms in high concentrations, and therefore this defense mechanism is generally referred to as 'quantitative'. A switch from the phenolic to the nitrogen pathway is frequently associated with the development of herbaceous growth-forms. Because many herbaceous plants are short-lived, the production of large amounts of defense agents is not economic because of the metabolic cost. Therefore, the production of smaller amounts of more toxic feeding deterrents, such as alkaloids and cyanogens, is favourable in these plants ('qualitative' defence strategy) (Lebreton, 1982). An additional advantage of alkaloid biosynthesis is the re-use of nitrogen incorporated in these micromolecules after the toxin is no longer needed as a feeding deterrent.

It has also been suggested that evolution of plant defences is associated with plant apparancy (Whittaker & Feeny, 1971; Cates & Rhoades, 1977) or resource availability in the growth environment (Coley & al., 1985). The latter authors demonstrated that slow-growing plants in nutrient-poor habitats prefer slow turnover rates of defensive chemicals (immobile defence strategy), whereas special feeding deterrents with fast turnover rates are favourable to fast growing plants from resource-rich habitats (mobile defence strategy). Although all these theories predict the type of plant defence, they do not explain the diversification of secondary metabolism as pointed out by Gottlieb & al. (1993).

An important aspect of protection against herbivory is the localization of toxins in the plant. In many studies toxins have been found at or near the leaf surface, where they are immediately effective after an attack by herbivores. In Asteraceae, for example, sesquiterpene lactones are often located in trichomes (Rodriguez & al., 1984). Epicuticular wax layers frequently contain terpenoids and phenolics in addition to alkanes and wax esters (Wollenweber, 1990). It should be mentioned that plant waxes also have physiological functions, e.g. the conservation of water, and protection against wind, rain, frost, and excessive sunlight (Martin & Juniper, 1970). Defensive phytochemicals are also important components of latices, which are liberated when the plant is injured. Well-known examples are the toxic phorbol esters from latices of Euphorbiaceae, and the bitter-tasting sesquiterpene lactones from the milky juice of chicory roots (*Cichorium intybus* L.).

The use of chemical characters in plant taxonomy has been a relatively unknown discipline until the early 1960s. Especially in the last few decades a wealth of information about the distribution of secondary metabolites has become available, and phytochemistry was established as an important tool to the taxonomist. This field of plant science is generally referred to as chemotaxonomy, chemosystematics, (bio)chemical systematics, or micromolecular systematics (Bisby & al., 1980; Harborne & Turner, 1984; Hegnauer, 1986; Hegnauer, 1962-1994).

Although chemical characters have proved to be useful in many taxonomic studies, chemotaxonomists have to deal with the same problems as biosystematists. First, a given chemical compound may be present in several taxa as a result of common ancestry (homology) or due to parallel or convergent evolution (non-homologous), viz. the development of a similar character state (chemical compound) in different lineages. Furthermore, secondary compounds may be formed via different biogenetic pathways. Thus, the presence of similar chemical patterns in different taxa does not necessarily imply kinship. Secondly, if one attempts to infer directional (evolutionary) information from chemical data, it is difficult to decide whether a secondary plant product represents the primitive or the derived condition. The subject of phytochemical evolution has been extensively studied by Gottlieb (1982a,b), who postulated that the complexity of secondary plant products tends to increase during evolution. Undoubtedly, this principle holds true in many taxa, but the possibility of reverse mutations should always be kept in mind. As chemical data do not, in principle, differ from comparative data from other sources, they can be examined by numerical methods to recognize patterns. These methods treat results in a phenetic way, which means that the objects (taxa) are grouped on the basis of overall similarity (Stuessy, 1990). On the other hand, few examples of handling chemical data in a phyletic way by cladistic analysis are available, probably because of the difficulty to assess the direction of biogenetic transformations. In addition, the low number of suitable characters found in most phytochemical studies in comparison to morphological or macromolecular studies discourage chemotaxonomists to use cladistic methods. When using any of the abovementioned methods, the chemotaxonomist has to cope with the problem of quantitative variation as well. It is very often difficult to decide wether a compound is present or not, especially when it is present in trace amounts only. The chromatographic techniques available nowadays enable the chemotaxonomist to detect secondary plant products in very low concentrations, but the question remains whether trace amounts are taxonomically significant, since trace compounds may be the result of 'bleeding' of intermediate products into other biogenetic pathways (Waterman & Gray, 1987).

The Eurasian Crassulaceae have been reported to contain alkaloids, tannins, flavonoids, cyanogenic glycosides, and epicuticular wax constituents. A compilation of the more systematically important phytochemical data from the literature is presented in

Table 1. Secondary metabolites from species of the Sedoideae and Sempervivoideae

Alkaloids	Source	Reference
(nor)hygrine, (nor)hygroline	*Sedum farinosum*	Stevens et al. (1992)
-	*Sedum aetnense*	-
-	*Sedum melanantherum*	-
-	*Sedum* ser. *Alpestria*	Stevens et al. (1993)
isopelletierine	*Sedum acre*	Schöpf and Unger (1957)
sedridine	-	Schöpf and Unger (1956)
-	*Sedum maximum*	Logar et al. (1974)
N-methylisopelletierine	*Sedum sarmentosum*	Marion and Chaput (1949)
-	*Sedum aizoon*	Krasnov et al. (1977)
-	*Sedum hybridum*	-
-	*Sedum purpureum*	-
N-methylallosedridine	*Sedum sarmentosum*	Marion and Chaput (1949), Beyerman et al. (1972)
pelletierine, sedridine and/or their N-methyl deriv.	*Sedum* ser. *Macaronesia*	Stevens et al. (1992)
-	*Sedum aetnense*	-
-	*Sedum anglicum*	-
-	*Sedum melanantherum*	-
-	*Sedum* ser. *Alpestria*	Stevens et al. (1993)
1-(2-piperidyl)-butan-2-one/ol and/or their N-methyl deriv.	*Sedum* ser. *Macaronesia*	Stevens et al. (1992)
1-(2-piperidyl)-pentan-2-one/ol and/or their N-methyl deriv.	-	Stevens et al. (1992)
sedamine	*Sedum acre*	Marion (1945)
-	-	Marion et al. (1951)
-	*Sedum maximum*	Logar et al. (1974)
-	*Sedum aizoon*	Krasnov et al. (1977)
-	*Sedum hybridum*	-
-	*Sedum purpureum*	-
-	*Sedum lydium*	Diak and Kohlmünzer (1981)
-, and related bases	*Sedum* ser. *Macaronesia*	Stevens et al. (1992)
hydroxysedamine	*Sedum acre*	Kooy (1976)
4-hydroxy(allo)sedamine	-	Halin et al. (1985)
3-hydroxy(nor)allosedamine and 5-hydroxysedamine	-	Ibebeke-Bomangwa and Hootelé (1987)
sedacrine	-	Nordal (1946)
-	-	Colau and Hootelé (1983)
-	*Sedum lydium*	Diak and Kohlmünzer (1981)
sedinine	*Sedum acre*	Franck (1958), Hootelé et al. (1980b)
-	*Sedum maximum*	Logar et al. (1974)
-	*Sedum aizoon*	Krasnov et al. (1977)
-	*Sedum hybridum*	-
-	*Sedum purpureum*	-

Table 1. Although I do not claim that our data are complete, this table may give the reader a good impression of the work that has been done in this field.

Table 1. (continued)

dihydrosedinine	*Sedum acre*	Colau and Hootelé (1983)
sedinone	-	Franck (1958)
-	-	Colau and Hootelé (1983)
-	*Sedum lydium*	Diak and Kohlmünzer (1981)
sedacryptine	*Sedum acre*	Hootelé et al. (1980a)
sederine	-	Hootelé et al. (1979)
hydroxysedinone	-	Kooy (1976)
lelobanidine	-	Francis and Francis (1977)
8-propyl-10-phenyllobeliolone	-	-
lobelanidine	-	-
hydroxysedinol	-	-
hydroxylelobanidine	-	-
lobelanidine glycoside	-	Piccinni-Leopardi et al. (1987)
nicotine	-	Marion (1945)
-	-	Gill et al. (1979)
-	*Sedum album*	-
-	*Sedum carpaticum*	-
-	*Sedum oppositifolium*	-
-	*Sedum pallidum*	-
-	*Sedum populifolium*	-
-	*Sedum telephium*	-
N-methylanabasine	*Sedum acre*	van der Wal et al. (1981)
sediene and sediendione	-	Maksimovic et al. (1990)
Tannins		
proanthocyanidins	*Sedum* spp.	Combier and Jay (1967)
-	*Sempervivum* spp.	-
-	*Aeonium arboreum*	-
Flavonoids		
kaempferol (and quercetin)	*Sempervivum ruthenicum*	Gumenyuk et al. (1970)
-	*Sempervivum* (5 spp.)	Combier and Jay (1967)
kaempferol glycosides	*Sedum crassipes*	Khetwal et al. (1988)
-	*Sedum* sect. *Gormania*	Denton and Kerwin (1980)
-	*Sedum middendorffianum*	Shnyakina and Zapesochnaja (1973a)
-	*Sedum pallescens*	Shnyakina and Zapesochnaja (1973b)
-	*Sedum sediforme*	Sakar et al. (1993)
-	*Sedum* sect. *Ternata*	Calie (1981)
-	*Sempervivum ruthenicum*	Gumenyuk et al. (1971)
herbacetin glycosides	*Sedum album*	Wolbis (1989)
sexangularetin glycosides	*Sedum sexangulare*	Combier et al. (1968)
-	-	Mnajed et al. (1972)
-	-	Wolbis and Kawalec (1990)

The species of *Sedum*, *Sempervivum*, and *Aeonium* included in our chemotaxonomic studies are given in Table 2. The species of *Sedum* and *Aeonium* are arranged according to infrageneric classifications by 't Hart (1991) and Mes (this volume), respectively. A few species from other genera of the Crassulaceae that have been examined for al-

Table 1. (continued)

quercetin glycosides	*Sedum acre*	Wolbis and Krolikowska (1988)
-	*Sedum album*	Wolbis (1989)
-	*Sedum middendorffianum*	Shnyakina and Zapesochnaja (1973a)
-	*Sedum* ser. *Rupestria*	Niemann et al. (1976)
-	*Sedum sediforme*	Sakar et al. (1993)
-	*Sedum* sect. *Ternata*	Calie (1981)
-	*Sempervivum ruthenicum*	Gumenyuk et al. (1972; 1982)
isorhamnetin glycosides	*Sedum acre*	Krolikowska (1966; 1972)
-	-	Wolbis and Krolikowska (1988)
-	*Sedum album*	Wolbis (1989)
-	*Sedum* ser. *Rupestria*	Niemann et al. (1976)
-	*Sempervivum ruthenicum*	Gumenyuk et al. (1972; 1982)
gossypetin glycosides	*Sedum album*	Wolbis (1989)
corniculatusin glycosides	*Sedum sexangulare*	Wolbis and Kawalec (1990)
-	*Sedum* ser. *Rupestria*	Niemann et al. (1976)
limocitrin (glycosides)	*Sedum acre*	Nordal (1946), Malterud and Nordal (1991)
-	-	Krolikowska (1966; 1972)
-	-	Wolbis and Krolikowska (1988)
-	*Sedum sediforme*	Sakar et al. (1993)
-	*Sedum sexangulare*	Wolbis and Kawalec (1990)
-	*Sedum* ser. *Rupestria*	Niemann et al. (1976)
myricetin glycosides	*Sedum middendorffianum*	Shnyakina and Zapesochnaja (1973a)
-	*Sedum montanum*	Stevens et al. (1994c)
-	*Sedum* ser. *Rupestria*	Niemann et al. (1976)
-	*Sedum sediforme*	Frigot (1960)
-	-	Sakar et al. (1993)
-, and 8-methoxy derivative	*Sedum sexangulare*	Wolbis and Kawalec (1990)
laricitrin (glycosides)	*Sedum crassipes*	Khetwal et al. (1988)
-	*Sedum* sect. *Gormania*	Denton and Kerwin (1980)
apigenin glycosides	*Sedum* sect. *Ternata*	Calie (1981)
luteolin glycosides	*Sedum* sect. *Ternata*	Calie (1981)
scutellarein glycosides	*Sempervivum ruthenicum*	Gumenyuk et al. (1972; 1982)
exudate flavonols		
-; que-3,7-dimethylether	*Aeonium manriqueorum*	Borges del Castello et al. (1968)
-; kae-3,4'-dimethylether	*Aeonium saundersii*	Wollenweber (1990)
-; kae-, 6-hydroxykae-, que-, and myr-methylethers	*Aeonium* (16 spp.)	Stevens et al. (1995b)
Terpenoids (epicuticular)		
diterpenoids (labdanediol)	*Aeonium lindleyi*	Baker et al. (1962)
pentacyclic triterpenoids	*Sedum* (19 spp.)	Stevens et al. (1994a)
-; triterpenol formates	*Sedum* ser. *Rupestria*	Stevens et al. (1994b)

kaloids and tannins are also included in this table. In the following I will present a synopsis of the chemotaxonomic utility of alkaloids, tannins, flavonoids, and epicuticular alkanes and terpenoids, which is based on both published and unpublished work.

Table 1. (continued)

Miscellaneous		
cyanogens (sarmentosin epox.)	*Sedum cepaea*	Nahrstedt et al. (1982)
- (sarmentosin)	*Sedum sarmentosum*	Fang et al. (1979)
alkanes	*Aeonium* spp.	Eglinton et al. (1962)
-	*Greenovia* spp.	-
-	*Monanthes* spp.	-
-	*Sedum anglicum*	-
-	*Sempervivum* spp.	-
-	*Sedum hispanicum*	Herbin and Robins (1968)
-	*Sedum pachyphyllum*	-
-	*Sedum lanceolatum*	Bowman (1983)
-, and other wax components	*Sedum* (30 spp.)	Stevens et al. (1994a)
α,ß-unsaturated-γ-lactones	*Sedum telephium* roots	Fung et al. (1990)

Meaning of '-' in the table: repeated entry from previous line.

Alkaloids of the Eurasian Sedoideae and Sempervivoideae

Introduction - Alkaloids have always fascinated mankind due to their wide range of pharmacological activities (see e.g. Mann, 1994), and the alkaloids of *Sedum* are no exception in this respect. The first *Sedum* alkaloid was isolated by Kolesnikow & Scharzmann in 1939, who named their substance 'sedamine' which probably consisted of a mixture of alkaloids (Franck, 1958). Marion (1945-1951) isolated another alkaloid from Canadian material of *S. acre,* which he also called 'sedamine'. However, Schöpf & Unger (1956) could not find sedamine in plant material of *S. acre* collected near Darmstadt, but isolated and identified another piperidine alkaloid, 2-piperidylpropan-2-ol, which they named 'sedridine'. These results led these authors to suggest the existence of 'chemical races' of *S. acre.* The absolute configuration of both alkaloids, sedamine and sedridine, was determined by Beyerman and co-workers (Beyerman & al., 1957).

Franck (1958) isolated four alkaloids from *Sedum acre,* two of which were new compounds. He proposed the names 'sedinone' and 'sedinine' for these compounds (Fig. 1). Franck also examined a number of other *Sedum* species for the presence of alkaloids, and found (traces of) alkaloids in *S. album, S. aizoon, S. crassipes, S. floriferum, S. rupestre, S. sexangulare* (= *S. mite*), *S. spurium,* and *S. telephium* (= *S. maximum*). He made no efforts, however, to identify the compounds, and I doubt that all these species contain piperidine alkaloids (Stevens & al., 1992). Also, the presence of alkaloids reported from *S. telephium* (= *S. maximum*), *S. aizoon,* and *S. hybridum* (Logar & al., 1974; Krasnov & al., 1977) could not be confirmed by Diak & Kohlmünzer (1981).

Kooy (1976) isolated two hydroxypiperidines, i.e. hydroxysedamine and hydroxysedinone, but was unable to determine the position of the hydroxy group. A number of 2,6-disubstituted piperidines similar to those previously reported from *Lobelia inflata* were identified by Francis & Francis (1977). Hootelé and co-workers (see Table 1) made an enormous contribution to the alkaloid chemistry of *Sedum acre.* Be-

Table 2. Classification* of the Crassulaceae included in our chemotaxonomic studies

Subfamily Genera Sections Series and species	Taxa examined for the presence of ALK	 TAN	 FLA	 WAX†
Crassuloideae				
Crassula L.				
C. multicava Lem.	+	+	-	-
Echeverioideae				
Echeveria DC.				
E. venezuelensis Rose	+	+	-	-
Pachyphytum Link, Klotzsch & Otto				
P. compactum Rose	+	+	-	-
Kalanchoideae				
Kalanchoe Adanson				
K. densiflora Rolfe, *K. synsepala* Baker	+	+	-	-
Bryophyllum Salisb.				
B. daigremontianum Berger	+	+	-	-
Sedoideae				
Sedum L.				
S. sect. *Africana* Fröd.				
S. meyeri-johannis Engler	+	+	+	+
S. sect. *Americana* Fröd.				
S. bourgaei Hemsley, *S. dendroideum* Moç. & Sessé, *S. reptans* Clausen	+	+	-	-
S. sect. *Sedum*				
S. ser. *Acria* Berger: *S. acre* L.	+	+	+	+
S. ser. *Macaronesia* Fröd.: *S. brissemoreti* Hamet, *S. farinosum* Lowe, *S. fusiforme* Lowe, *S. lancerottense* Murray, *S. nudum* Aiton	+	+	+	+
S. ser. *Macrosepala* (Reg. & Schm.) Boriss.: *S. aetnense* Tin.	+	-	-	-
S. ser. *Anglica* 't Hart: *S. anglicum* Huds.	+	+	+	+
S. ser. *Melananthera* 't Hart: *S. melanantherum* DC.	+	+	+	+

Table 2. (continued)

S. ser. *Alpestria* Berger: *S. alpestre* Vill., *A. annuum* L., *S. apoleipon* 't Hart, *S. borissovae* Balk, *S. euxinum* 't Hart & Alpinar, *S. grisebachii* Boiss. & Heldr., *S. laconicum* Boiss. & Heldr., *S. multiceps* Coss. & Dur., *S. sexangulare* L., *S. tuberiferum* Stoj. & Stef., *S. tuberosum* Coss. & Let., *S. ursi* 't Hart, *S. urvillei* DC.	+	(+)	+	+
S. ser. *Samia* 't Hart: *S. samium* Runemark & Greuter	+	+	–	–
S. ser. *Litorea* 't Hart: *S. litoreum* Guss.	+	+	–	+
S. ser. *Alba* Berger: *S. album* L.	+	+	+	+
S. ser. *Atrata* 't Hart: *S. atratum* L.	–	–	+	+
S. ser. *Dasyphylla* 't Hart: *S. dasyphyllum* L., *S. granatense* Pau, *S. mucizonia* (Ortega) Hamet	–	–	–	+
S. ser. *Lydia* 't Hart: *S. lydium* Boiss.	+	+	+	+
S. ser. *Pedicellata* 't Hart: *S. andegavense* (Cav.) DC., *S. brevifolium* DC., *S. pedicellatum* Boiss. & Reuter, *S. candollei* Hamet	–	–	–	+
S. ser. *Propontica* Berger: *S. stellatum* L.	+	+	–	–
S. ser. *Rupestria* Berger: *S. amplexicaule* DC., *S. pruinatum* Brot., *S. forsterianum* Sm., *S. montanum* Song. & Perr., *S. ochroleucum* Chaix, *S. rupestre* L., *S. sediforme* Pau	+	+	+	+
S. ser. *Monanthoidea* (Batt.) Fröd.: *S. jaccardianum* Maire & Wilczek, *S. modestum* Ball	–	–	+	+
S. ser. *Pubescens* T. Mes: *S. pubescens* Vahl.	–	–	+	+
Sempervivoideae				
Aeonium Webb & Berth.				
A. sect. *Aeonium*				
A. arboreum (L.) W. & B., *A. gorgoneum* J.A. Schmidt, *A. leucoblepharum* Webb ex A. Rich., *A. holochrysum* W. & B., *A. rubrolineatum* Svent., *A. simsii* (Sweet) Stearn	(+)	(+)	+	+
A. sect. *Anodontonium* Berger				
A. lindleyi W. & B., *A. goochiae* W. & B., *A. viscatum* Bolle	–	–	+	+
A. sect. *Canariense* (Christ) Praeger				
A. canariense (L.) W. & B., *A. cuneatum* W. & B., *A. palmense* Webb ex Christ, *A. subplanum* Praeger, *A. tabuliforme* (Haw.) W. & B., *A. virgineum* Webb ex Christ	–	–	+	+
A. sect. *Chrysocome* Webb ex Christ				
A. smithii (Sims) W. & B., *A. spathulatum* (Hornem.) Praeger	–	–	+	+

Table 2. (continued)

A. sect. *Leuconium* Berger				
A. castello-paivae Bolle, *A. ciliatum* Webb & Berth., *A. decorum* Webb ex Bolle, *A. gomerense* (Praeger) Praeger, *A. haworthii* Salm-Dyck ex W. & B., *A. hierrense* (Murr.) Pit. & Proust, *A. lancerottense* (Praeger) Praeger, *A. nobile* (Praeger) Praeger, *A. percarneum* (Murray) Pit. & Proust, *A. urbicum* (Smith ex Hornem.) W. & B., *A. valverdense* (Praeger) Praeger	(+)	(+)	+	+
A. sect. *Patinaria* (Lowe) Berger				
A. glandulosum (Aiton) W. & B.	–	–	+	+
A. sect. *Petrothamnium* (Webb ex Christ) Liu				
A. saundersii Bolle, *A. sedifolium* (Webb ex Bolle) Pit. & Proust	(+)	(+)	+	+
A. sect. *Pittonium* Berger				
A. glutinosum (Aiton) W. & B.	+	+	+	+
Jovibarba Opiz				
J. heuffelii (Schott) A. & D. Löve	+	–	+	+
Sempervivum L.				
S. x adenotrichum, *S. altum* Turrill, *S. arachnoideum* L., *S. armenum* Boiss. & Huet, *S. atlanticum* Ball, *S. balsii* Wale, *S. borissovae* Wale, *S. calcareum* Jordan, *S. cantabricum* Huber, *S. caucasicum* Ruprecht, *S. charadzeae* Gurgenidze, *S. ciliosum* Craib, *S. dzhavachischvili* Gurgenidze, *S. erythraeum* Velen., *S. furseorum* Muirhead, *S. grandiflorum* Haw., *S. x guiseppii*, *S. ingwersenii* Wale, *S. italicum* Ricci, *S. kindingeri* Adamovic, *S. kosaninii* Praeger, *S. leucanthum* Pancic, *S. macedonicum* Praeger, *S. marmoreum* Griseb., *S. montanum* L., *S. nevadense* Wale, *S. ossetiense* Wale, *S. pumilium* Bieb., *S. reginae-amaliae* Halàcsy, *S. ruthenicum* Schnitt. & Lehmann, *S. tectorum* L., *S. thompsonianum* Wale, *S. transcaucasicum* Muirhead, *S. zeleborii* Schott	(+)	(+)	+	+

* Infrageneric classification of *Sedum* and *Aeonium* according to 't Hart (1991) and Mes (1995), respectively.

† Key: ALK, alkaloids; TAN, tannins; FLA, flavonoids; WAX, wax components. Signs: – not examined, (+) at least one species from the genus, section or series examined, + (almost) all species of the genus, section or series examined.

sides the principal alkaloids, they isolated a range of minor alkaloids not previously known, and established their structure and absolute configuration. Recently, Driessens & Hootelé (1991) described the total synthesis of sedinone and sedacrine by application of anodic oxidation.

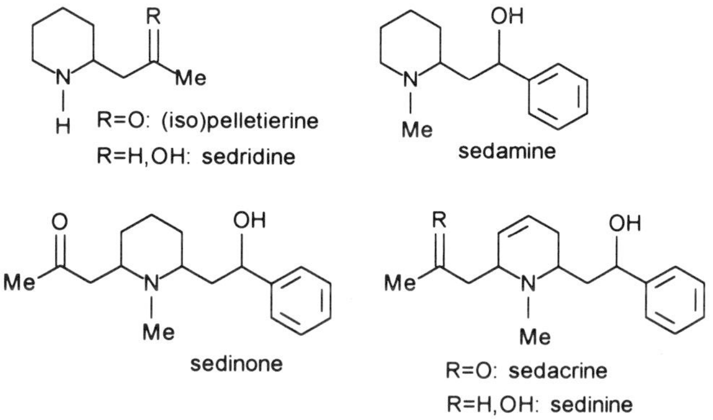

Figure 1. Alkaloids from *Sedum acre.*

Besides their occurrence in the Crassulaceae, *Sedum* alkaloids have been found in a few other families of higher plants only. Pelletierine occurs in the pomegranate plant (*Punica granatum*, Punicaceae), and in *Duboisia myoporoides* (Solanaceae). Its N-methyl derivative has been reported from *Punica granatum* and *Lupinus formosus* (Leguminosae). The pyrrolidine alkaloids, hygrine and hygroline, which we detected in several species of *Sedum* (see below), are also intermediate products in the biosynthesis of tropane alkaloids, and are thus commonly encountered in families containing tropane alkaloids, e.g. in Solanaceae, Erythroxylaceae, and Convolvulaceae. Several 2-monosubstituted and 2,6-disubstituted piperidine alkaloids have been found in species of Campanulaceae (*Lobelia inflata*, *L. hassleri*, *L. nicotianaefolia*, *Isotoma longiflora*, *Laurentia longiflora*, and *Campanula medium*) (Hegnauer, 1989; Harborne & Baxter, 1993). However, the typical *Sedum acre* alkaloids, i.e. sedamine, sedacrine, sedinine, and sedinone, have not been reported from other families to our knowledge.

Biosynthesis of pyrrolidine and piperidine alkaloids from Crassulaceae - The pyrrolidine alkaloid norhygrine and its piperidine analogue pelletierine have been shown to share the same biosynthetic route. The pyrrolidine and the piperidine nuclei originate from ornithine and lysine, respectively. In the case of piperidine alkaloids, decarboxylation and transamination of lysine leads to the formation of 5-aminopentanal, which is in equilibrium with the cyclic imine Δ^1-piperideine (Mann, 1987). This active intermediate may react with beta-keto acids via a Mannich type condensation, and subsequent decarboxylation yields pelletierine (Gupta & Spencer, 1969; Mann, 1987) (Fig. 2). After formation of the skeleton, modifications such as N-methylation and hydrogenation of the side-chain keto group may occur. The N-methyl groups are derived from S-adenosyl methionine or methanoate (Mann, 1987). The side-chain of sedamine has been found to originate from phenylalanine (Gupta & Spencer, 1967).

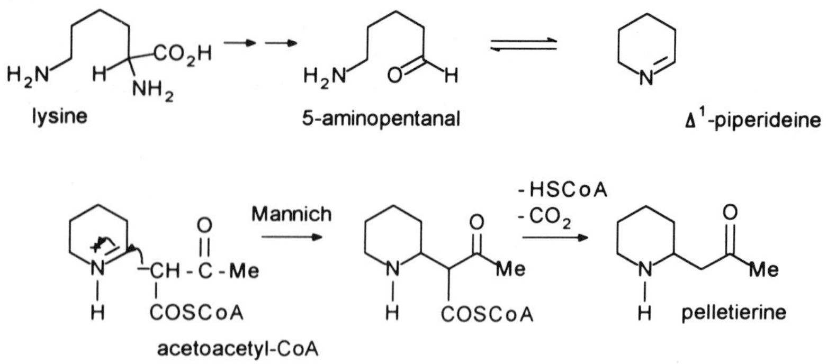

Figure 2. Biosynthesis of pelletierine (after Mann, 1987).

Distribution of alkaloids in the Crassulaceae - 53 species of Crassulaceae were examined for the presence of alkaloids. Only species of *Sedum* sect. *Sedum* were found to contain alkaloids, which were identified as pyrrolidines and piperidines. Species of *Aeonium, Bryophyllum, Crassula, Greenovia, Kalanchoe, Sedum* sect. *Africana, S.* sect. *Americana,* and *Sempervivum* did not contain alkaloids. This distribution pattern is in good agreement with a phylogeny of the Crassulaceae based on chloroplast DNA restriction-site variation (Fig. 3) (van Ham, 1994). Within *Sedum* sect. *Sedum* the presence of alkaloids is restricted to the so-called *Acre*-clade, which includes the species with basally free or spurred sepals and a reticulate testa. Since the pyrrolidines and piperidines from species of *Sedum* are rare in nature, the biosynthetic gain to produce these alkaloids has probably occurred once in the family at the very base of the *Acre*-clade. In this view, the absence of alkaloids in *S.* sect. *Africana, S.* sect. *Americana,* and *Pachyphytum* must be the result of subsequent biosynthetic loss. That this phylogenetic tree has considerable predictive value in the hunt for alkaloids is exemplified by the presence of two piperidine alkaloids, i.e. pelletierine and N-methylpelletierine, in the *Acre*-clade species *Echeveria venezuelensis.*

Furthermore, Diak & Kohlmünzer (1981) reported the occurrence of piperidines in *Sedum lydium* (see also Harborne & Baxter, 1993). This species does not belong to the *Acre*-clade (van Ham, 1994), however, and we could not confirm the presence of alkaloids in a plant originating from the centre of distribution of this species (Western Anatolia).

The alkaloid variation within the *Acre*-clade is by and large in agreement with the classification of the species into series based on biosystematic data ('t Hart, 1991; Stevens & al., 1993). *Sedum acre* (the only species of *S.* series *Acria*) contains the widest variety of piperidine alkaloids with about 20 alkaloids. Sedridine, sedamine, 'hydroxysedamine', sedacrine, sedinine, and sedinone are the major alkaloids of this species. *S.* series *Macaronesia* is characterized by the presence of 2-monosubsituted piperidines bearing propan-2-one (pelletierine), propan-2-ol (sedridine), butan-2-one, butan-2-ol, pentan-2-one, and pentan-2-ol side-chains, as well as their N-methyl derivatives (Fig. 4) and derivatives of sedamine. The alkaloid patterns of *S.* series *Macrosepala, S.* series *Anglica, S.* series *Melananthera, S.* series *Alpestria, S.* series *Litorea,* and *S.* series *Samia* are similar. Pyrrolidine alkaloids in addition to piperidine alkaloids of the pelletierine type are commonly encountered in these taxa. *S. farinosum* of *S.* series *Macaronesia* also shares this pattern, and therefore the classification of this species in another series has been advocated (Stevens & al., 1992).

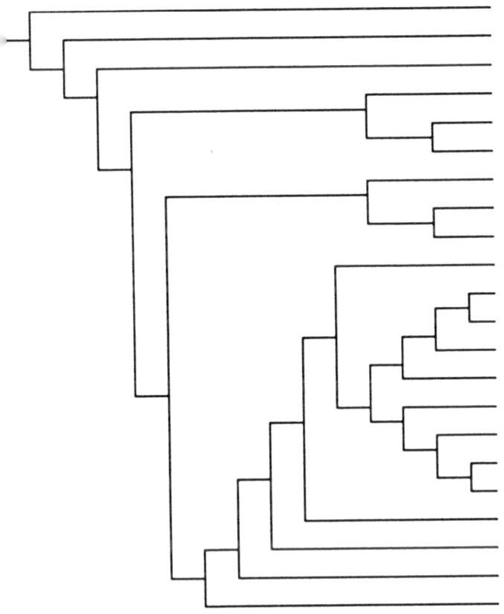

	1	2	3	4	5	6
Crassula multicava		●			●	●
Kalanchoe			●		x	x
Sedum stellatum					●	●
Sempervivum					●	●
Sedum ser. Rupestria					●	●
Aeonium			●		●	●
Sedum brevifolium					●	?
Sedum album					●	
Sedum lydium					●	
Sedum bourgaei	●	●	●		●	●
Sedum dendroideum			●		●	
Echeveria venezuelensis			●	●	●	●
Pachyphytum compactum					●	●
Sedum reptans		●			●	●
Sedum fusiforme	●	●	●	●		
Sedum farinosum	●	●		●	●	●
Sedum anglicum	●	●		●	●	●
Sedum melanantherum	●	●		●		
Sedum litoreum				●	●	?
Sedum ser. Alpestria		●		●	x	
Sedum meyeri-johannis	●	●	●		●	
Sedum acre	●	●		●		

Figure 3. Distribution of three morphological and three phytochemical characters plotted against a phylogeny of the Crassulaceae based on chloroplast DNA restriction-site variation (van Ham, 1994). Character numbers: (1) sepals basally free or spurred (versus connate); (2) testa reticulate (versus costate or bipapillate); (3) shoot woody (versus herbaceous); (4) piperidine and/or pyrrolidine alkaloids present (versus absent); (5) proanthocyanidins present (versus absent); (6) galloyl esters present (versus absent). A dot indicates presence of the character; x indicates presence of the character in the majoritiy of the species; ? indicates that the character was not determined.

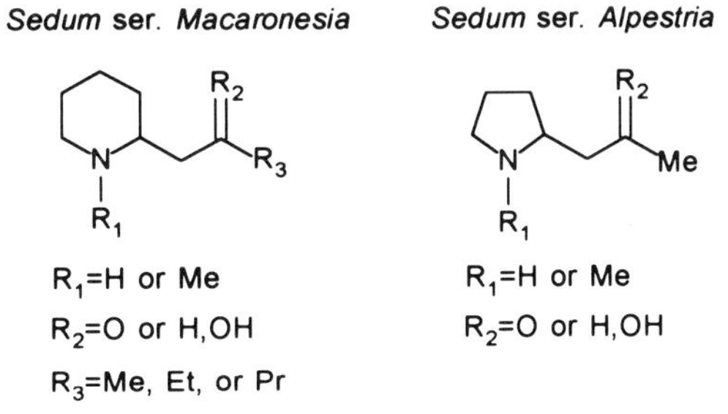

Figure 4. Some alkaloids from species belonging to the *Acre*-clade

Tannins of the Crassulaceae

Chemistry of condensed tannins and galloyl esters - Vegetable tannins are a group of polyphenolic compounds with adstringent properties. They form complexes with macromolecules, such as proteins and polysaccharides, and alkaloids. Their most important function in plants is probably to make plant tissues unpalatable to herbivores (Harborne, 1988), but tannin concentrations should exceed normal dietary levels in order to be effective (Grayer, 1992).

Vegetable tannins can be divided into two major chemical groups: condensed tannins and hydrolysable tannins. The condensed tannins or proanthocyanidins are dimers, oligomers, or polymers of flavan-3-ol units, which are predominantly 4→8 or 4→6 linked with each other (Fig. 5). The biosynthesis of the monomeric flavonoid units goes via the normal flavonoid pathway. Hydroxylation of flavanone precursors at C-3, and subsequent reduction steps yield the flavan-3-ols, i.e. (epi)catechin and (epi)gallocatechin. Catechin and epicatechin condense to form procyanidins, while (epi)gallocatechin units polymerize into the prodelphinidins (Haslam, 1989; Porter, 1989). Although pure polymers of either of the two monomers are found in nature, copolymers in which either procyanidin or prodelphinidin predominates are far more common (Porter & al., 1986; Porter, 1989).

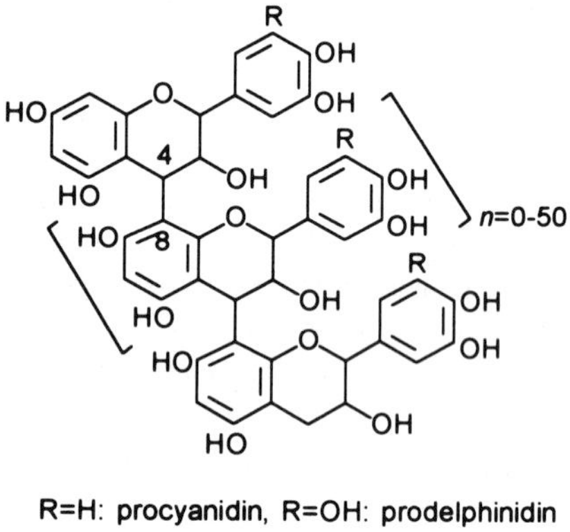

Figure 5. Structure of the proanthocyanidins.

Condensed tannins are converted into red-coloured anthocyanidins on treatment with hot alcoholic solutions of mineral acids, which explains why condensed tannins are also called 'proanthocyanidins'. The acid-catalyzed cleavage of the oligo and polymers into monomeric anthocyanidins is often employed to detect condensed tannins in plant tissues. In older literature reports one may find the term 'leucoanthocyanidins' used as a synonym for the proanthocyanidins (Jay, 1970). Haslam (1982) defined leucoanthocyanidins as monomeric proanthocyanidins. More recently, Porter (1988) proposed to use the term for 'monomeric flavanoids which produce anthocyanidins by cleavage of a C-O bond on heating with mineral acid (including flavan-3,4-diols and flavan-4-ols)'. He also suggested to redefine the proanthocyanidins as 'all compounds which produce anthocyanidins by cleavage of a C-C bond'. Esters of gallic acid or galloyl esters represent a structurally diverse group of secondary metabolites. Gallic acid has been found in esterified combination with other gallic acid molecules (depsides), simple phenols (e.g. monogalloylarbutin), flavonoids (including proanthocyanidins), and glucose (e.g. glucogallin) (Harborne & Baxter, 1993). An important group of galloyl esters are the gallotannins which are depside derivatives of penta-O-galloyl-D-glucose. Ellagitannins differ from the gallotannins in the additional presence of hexahydroxydiphenic acid groups formed through oxidative coupling of two vicinal galloyl residues in the pentagalloyl glucose molecule. The gallotannins and ellagitannins together form the hydrolysable tannins (Haslam, 1989). On hydrolysis, galloyl esters give gallic acid while ellagitannins also release ellagic acid (di-lactone of hexahydroxydiphenic acid) (Fig. 6) which can be analysed with colorimetric and chromatographic techniques (Porter, 1989).

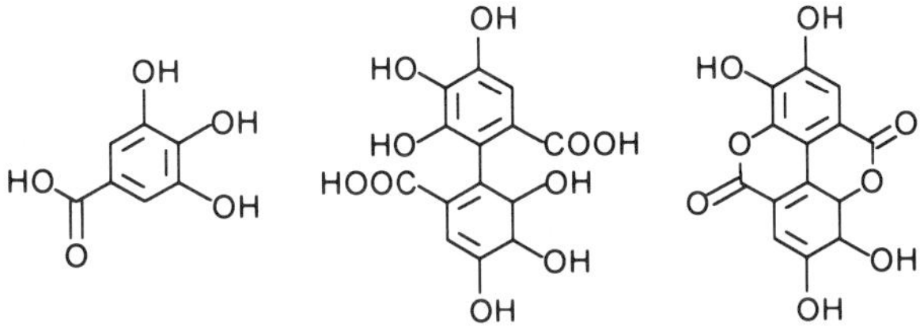

A B C

Figure 6. Structure of galloyl esters. 6A. Gallic acid. 6B. Hexahydrodiphenic acid. ellagic acid.

Distribution of tannins versus alkaloids in the Crassulaceae - Proanthocyanidins and galloyl esters of 36 species of Crassulaceae were qualitatively and quantitatively determined, and their distribution was compared with the alkaloid records of the species. The majority of the species contained considerable amounts of gallotannins and proanthocyanidins in which prodelphinidin predominates (Stevens & al., 1995a).

It has been postulated that evolution proceeds by the transition from general quantitative (e.g. tannins) to specific qualitative (e.g. alkaloids) defence agents (Whittaker & Feeny, 1971; Lebreton, 1982; Coley & al., 1985; Gottlieb & al., 1993). Although these general considerations apply to comparisons at familial and higher classificatory levels, a similar trend can also be observed at the generic and infrageneric levels in the Crassulaceae. The alkaloid-containing *Sedum* series *Acria*, *S.* series *Macaronesia*, *S.* series *Melananthera*, and *S.* series *Alpestria* either contain small amounts or are in want of tannins, whereas most taxa lacking alkaloids have high tannin levels. From a plot of the distribution of alkaloids and tannins against a phylogeny of the family (Fig. 3) one may conclude that the presence of alkaloids represents the derived rather than the primitive condition, which favours the above hypothesis. By contrast, the presence of alkaloids is strongly correlated with some floral and seed characters which are considered to be primitive ('t Hart & Koek-Noorman, 1989) (see Fig. 3). The most parsimonious explanation for this apparent contradiction seems to be a reversal at the very basis of the *Acre*-clade resulting in floral structures (free and spirally arranged sepals) similar to that of the primitive Magnolioids and Rosoids (van Ham, 1994; Stevens & al., 1995a).

Because the occurrence of proanthocyanidins is frequently associated with the woody habit of higher plants, the use of proanthocyanidins as chemical characters loses value in taxonomic studies (Waterman & Gray, 1987). In various angiosperm lineages the transition from woody to herbaceous growth-forms is often accompanied by loss of proanthocyanidins. The correlation between the presence of proanthocyanidins and woodiness, however, is completely absent in the Crassulaceae (Fig. 3). Most herbaceous Crassulaceae have retained the ability to accumulate proanthocyanidins, and in some woody species (*Sedum fusiforme* and *S. nudum* of the Macaronesian Sedoideae) the proanthocyanidins have even been replaced by alkaloids. These observations lend support to the hypothesis that the woody taxa of the Crassulaceae are derived from herbaceous ancestors ('t Hart & Koek-Noorman, 1989).

The galloyl esters are restricted to the dicotyledons, and, more specifically, their distribution is largely confined to the primitive members of the Hamamelididae-Dilleniidae-Rosidae complex. In addition, species belonging to these classes are rich sources of proanthocyanidins of the delphinidin type. As a member of the Hamamelididae-Dilleniidae-Rosidae complex, the Crassulaceae also contain both prodelphinidins and galloyl esters. The Crassulaceae differ from the related Saxifragaceae and Penthoraceae in the absence of ellagitannins, which thus seems to be a useful character at the familial level (Jay, 1970).

It has been suggested that the abundant presence of galloyl esters and condensed tannins makes the production of specific metabolites (qualitative defence agents) redundant (Gottlieb & al., 1993) from an ecological viewpoint. This is certainly in agreement with the observed alkaloid/tannin dichotomy in many taxa, including the Crassulaceae, but, in Gottlieb's vision, the availability of active oxygen rather than selection pressures determines chemical diversification. In view of the well-known antioxidant properties of tannins (radical scavenging), it seems possible that high subcellular concentrations of tannins inhibit active oxygen transfer (oxygenation) and thus chemical diversification. To the examples given by Gottlieb & al. (1993), the Crassulaceae could be added. On the whole, the Crassulaceae are poor in alkaloids, and, moreover, the chemical variation in the piperidine and pyrrolidine alkaloids originates from condensation reactions rather than from oxygenation. A similar situation is found in the terpenoids of the Eurasian Sedoideae and Sempervivoideae, where chemical variation is not the result of secondary oxygenation (see below).

Flavonoids of the Eurasian Sedoideae and Sempervivoideae

Introduction - Flavonoids are phenolic compounds which occur in all plants except for most algae. Their biosynthesis is of mixed origin: the B-ring and part of the heterocyclic ring are provided by cinnamic acid derivatives, while the A-ring of the flavonoid skeleton is formed from three malonyl-CoA units (see structure in Fig. 7). Subsequent condensation, reduction, hydroxylation, methylation, and glycosylation reactions give rise to the enormous structural variation known in the plant kingdom (for a recent review on the biosynthesis see Heller & Forkman, 1994). It should be noted that the abundant presence of galloyl esters seems to have little effect on the biosynthesis of cinnamate derived metabolites, such as flavonoids and lignoids (Gottlieb & al., 1993; see previous section). Presently, a total of ca 380 flavonols and 300 flavones with hydroxy and/or methoxy substitution, as well as some 1370 glycosides of these two classes are known (Wollenweber, 1994; Williams & Harborne, 1994). Whereas flavonoid glycosides are predominantly found inside the plant cell, flavonoid aglycones very often occur externally on leaf surfaces and in leaf resins (Wollenweber, 1994).

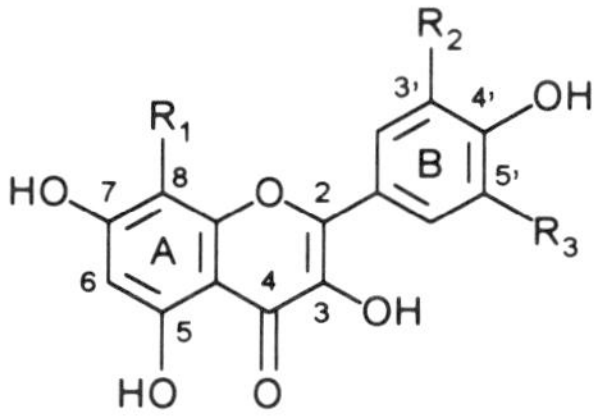

R_1	R_2	R_3	flavonoid
H	H	H	kaempferol
OH	H	H	herbacetin
OMe	H	H	sexangularetin
H	OH	H	quercetin
H	OMe	H	isorhamnetin
OH	OH	H	gossypetin
OMe	OH	H	comiculatusin
OMe	OMe	H	limocitrin
H	OH	OH	myricetin
H	OMe	OH	laricitrin
OH	OH	OH	hibiscetin

Figure 7. Flavonoids of Eurasian *Sedum* and *Sempervium.*

A variety of functions have been ascribed to the flavonoids. As pigments (anthocyanins) or copigments (flavones and flavonols) in floral structures they act as pollinator attractants. For example, the flavonol gossypetin provides the yellow flower colour of primroses. However, the yellow colour of many *Sedum* species is due to the presence of carotenoids, a group of tetraterpenes. Other functions of flavonoids include the protection against herbivore pests (especially in the case of exudate flavonoids and isoflavone accumulation induced by infections of pathogenic fungi and bacteria), the absorption of UV light (McClure, 1985), and the regulation of auxin transport (Jacobs & Rubery, 1988).

Flavonoids are popular chemical characters in chemosystematics because of their universal occurrence in higher plants and their great structural diversity. In addition, flavonoids are chemically stable and relatively easy to isolate, separate, and identify. Fresh as well as dried plant material (herbarium specimens) can be examined (Harborne & Turner, 1984), and it is even possible to detect flavonoids in fossils (Giannasi & Niklas, 1977). Most flavonoid work has been concerned with lower levels of hierarchy, and often flavonoid chemistry proved useful in solving problems of interspecific hybridization (see Harborne & Turner, 1984). Flavonoid characters can be treated in much the same way as morphological data by numerical methods, but one must be cautious with the construction of phylogenies from flavonoid data (Stuessy & Crawford, 1983).

Flavonoid variation in the Eurasian Sedoideae - The flavonoid aglycones of 29 species of *Sedum* sect. *Sedum* and one species of *S.* sect. *Africana* were examined in hydrolysed plant extracts of stems and leaves. In this initial survey, we gave low priority to the investigation of the complex glycosylation patterns for two main reasons. First, identification of (most of) the flavonoid glycosides from a large number of species would require extensive isolations, which would consume too much time. Secondly, we expected the glycosylation patterns to be less relevant as judged from preliminary work and available literature reports. For instance, the principal aglycones of *S.* series *Rupestria* are quercetin and myricetin, while the major glycosides are the 3-rhamnosides of these aglycones (Niemann & al., 1976; Stevens & al., 1994c). Further, 14 flavonol glycosides were reported from *S. acre* (Wolbis & Krolikowska, 1988), 9 from *S. sexangulare* (Wolbis & Kawalec, 1990), and 16 from *S. album* (Wolbis, 1989): these species contained similar glycosylation patterns but differed mainly in their aglycone composition. Other authors also chose to study flavonoid variation at the aglycone level before considering the complexity of the glycosides, e.g. Williams (1978), Stuessy & Crawford (1983), and Williams & al., (1986).

A total of 10 flavonoid aglycones were detected and quantified with gas chromatography. The flavonoids were identified as kaempferol, quercetin, isorhamnetin, myricetin, and their 8-hydroxy and/or 8-methoxy derivatives (Fig. 7). The majority of these aglycones were widely distributed in *Sedum*, probably resulting from a high degree of evolutionary parallelism. Consequently, attempts to infer phylogenetic information from a flavonoid data matrix of the usual present/absent format using approaches described by Humphries & Richardson (1980), Richardson & Young (1982), and Stuessy & Crawford (1983) were not successful. Bohm & al. (1984; and references cited) encountered a similar situation in the related Saxifrageae, in which the diversity of flavonoid biochemistry seems to reflect the widespread cytological and morphological parallelisms present in this taxon.

Principal components analysis of the quantitative flavonoid data also proved difficult because of the influence of the abundantly present flavonoids on the distribution of the plants in the scatter plot. Therefore, the original data were transformed in the following way. Absence was scored 0, traces (0.01 mg/g fresh weight) were given a value of 1, minor flavonoids (0.02-0.1 mg/g fresh weight) a value of 2, and major flavonoids (> 0.1 mg/g fresh weight) received a value of 3. The resulting scatter plot is presented in Figure 8. Four groups are separated from the bulk of the species situated near the centre of the scatter plot. *Sedum* series *Rupestria* forms the most distinct group. Without exception, the species of this series contain large amounts of quercetin and myricetin. The second group represents about half of *S.* series *Alpestria.* The species of this cluster are characterized by the abundant presence of the unusual flavonol sexangularetin, which was found for the first time in *S. sexangulare* of *S.* series *Alpestria* (Combier & al., 1968). Interestingly, the three *S. sexangulare* plants examined had very similar flavonoid patterns, whereas the distribution of alkaloids in the same specimens showed considerable intraspecific variation. By contrast, all species of *S.* series *Alpestria* were found to contain similar alkaloids, but the distribution of flavonoids shows a clear dichotomy (Fig. 8). As yet, these marked differences cannot be correlated with morphological and cytological features of this comparium ('t Hart, in press). The three plants of *S. album* were all grouped in the third cluster due to the presence of gossypetin and myricetin. The fourth group consists of the three plants of *S. brevifolium* included in the survey. Herbacetin and gossypetin are the principal flavonols of this species. Recently, a close relationship was found between *S. brevifolium* and *S. atratum* in a study of chloroplast DNA restriction-site variation in Eurasian *Sedum* (van Ham, 1994). The distinct flavonoid pattern of *S. brevifolium* encouraged us to examine *S. atratum* as well. *S. atratum* was also found to contain large amounts of both flavonols, which emphasizes the close evolutionary relationship between the two species.

The remainder of the species examined are situated as a diffuse cluster near the centre of the scatter plot (Fig. 8). It includes *S. acre* and species of *S.* series *Alpestria*, *S.* series *Anglica*, *S.* series *Macaronesia*, and *S.* series *Melananthera*. *S. acre* is characterized by the combination of isorhamnetin and limocitrin, while kaempferol and quercetin are the principal flavonols of most of the other taxa.

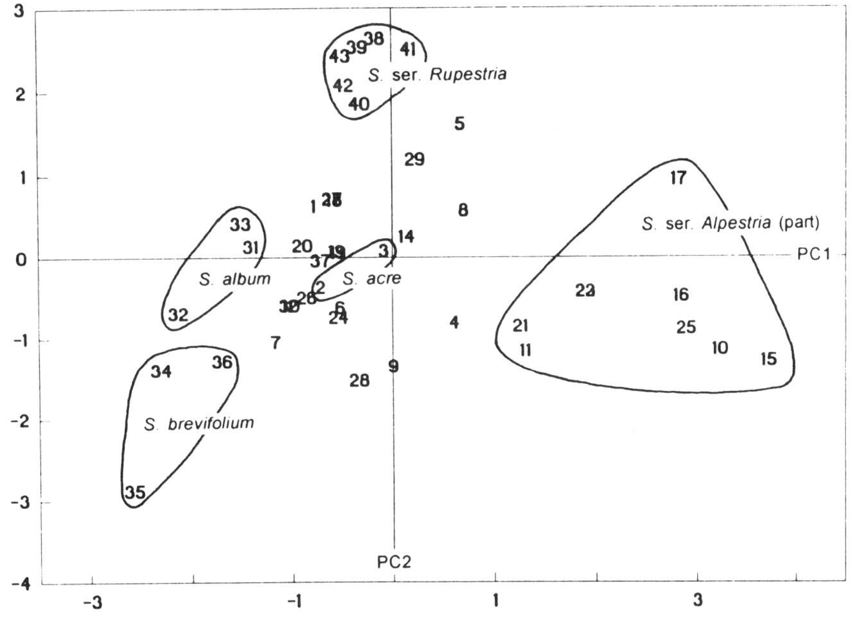

Figure 8. Principal components analysis of the flavonoid data of Eurasian *Sedum*. The first and second axis account for 27% and 20% of the variance, respectiveley

Flavonoid variation in the Sempervivoideae - 34 species of *Sempervivum* and one species of *Jovibarba*, i.e. *J. heuffelii*, were investigated for the presence of flavonoids using the same methods. *Sempervivum* showed remarkably little infrageneric variation. Four flavonols were detected, i.e. kaempferol, herbacetin, quercetin, and myricetin. Kaempferol was the principal flavonol of all species. Given the homogeneous flavonoid patterns of the species, it was not possible to distinguish infrageneric groups.

Taxonomic significance of alkane patterns in the Eurasian Sedoideae and Sempervivoideae - The epicuticular wax layers of Eurasian Sedoideae and Sempervivoideae contain alkanes, alcohols, fatty acids, fatty acid methylesters, aldehydes, alkyl esters, and terpenoids. Of the aliphatic constituents, only the alkane chain length distribution proved to be of considerable taxonomic value.

The alkane chain length distribution of 30 Eurasian species of *Sedum* was studied using multivariate data analysis methods. Cluster analysis distinguished three groups which mainly differed in the amounts of nonacosane, hentriacontane, and tritriacontane. *S. nudum*, *S. fusiforme*, and *S. brissemoreti* of *S.* series *Macaronesia* grouped in the first cluster. The second cluster contained *S. farinosum*, *S. melanantherum*, and *S. anglicum*. This clustering supports the exclusion of *S. farinosum* from *S.* series *Macaronesia* as proposed on the basis of alkaloid data (see above). The other two species with reticulate seeds and free sepals, i.e. *S. acre* and *S. meyeri-johannis*, were placed in the third cluster along with the remainder of the species included in the study. The arrangement of the species in this large cluster revealed no clear pattern, however (Stevens & al., 1994a).

The alkane patterns of *Sempervivum* were of little taxonomic value at the infrageneric level. The alkane chain length distributions of the species were very similar with tritriacontane being the dominant homologue.

By contrast, alkane variation in *Aeonium* is enormous. It has been studied by Eglinton & al. (1962), who found some trends in the alkane patterns which could be related to the sectional classification, but there was no clear distinction between the sections. We obtained similar results: with principal components analysis, four groups could be distinguished. Taxa with hentriacontane as dominant alkane were placed in the first group, which contained species from three different sections, including *A. glutinosum.* All species of *A.* sect. *Aeonium* and *A.* sect. *Leuconium* were placed in the second group (C_{33} dominant), together with some other species. The third and the fourth group were characterized by the presence of medium and high isoalkane proportions, and contained all species of *A.* sect. *Petrothamnium*, *A.* sect. *Goochia*, as well as a few other species, including *A. smithii* and *A. palmense.* The alkane patterns of two North African *Sedum* species, *S. jaccardianum* and *S. pubescens*, which are thought to be closely related to the Macaronesian Sempervivoideae (Mes, this volume) also show high isoalkane proportions, in particular *S. pubescens.* These results support the hypothesis of a common ancestor capable of producing isoalkanes, since a repeated gain of isoalkane production is not likely to have taken place.

From this clustering it can be learned that the general trend in alkane variation is indeed correlated with sectional classification, but 'mis-classifications' are also quite common. Moreover, infraspecific variation was considerable (by contrast, differences between duplicate analyses of the same plants - that is, same clone - were negligible), which may result in classification of plants of the same species in different groups (*Aeonium holochrysum* and *A. canariense*). This makes the alkanes less valuable as chemotaxonomic markers. Therefore, caution should be exercised in interpreting the results, especially when the number of plants of a species examined is small.

Taxonomic and evolutionary significance of epicuticular terpenoids in Eurasian *Sedum* and *Aeonium* - Epicuticular triterpenoids have been shown to cause glaucousness in *Dudleya* (Crassulaceae) (Manheim & Mulroy, 1978), and also in Eurasian *Sedum* (Stevens et al., 1994a; 1994b). The presence of large deposits of these compounds is limited to the blue- or grey-leaved plants (Fig. 9). Notwithstanding the discontinuous distribution of epicuticular triterpenoids in Eurasian *Sedum*, they proved to be of great taxonomic value at the infrageneric level (Table 4). Some 14 triterpenes were identified as being representatives of the ursane, oleanane, multiflorane, taraxerane, fernane, and lupane series (Stevens & al., 1994a) (Fig. 10). The presence of formic acid esters of the triterpenols, germanicol and fernenol, which we reported for the first time, is typical of *S.* series *Rupestria*, and emphasizes the micromolecular distinctiveness of this comparium. In addition to herbacetin and gossypetin (both flavonoids), multiflorenyl acetate (a triterpene) appeared to be a useful character to demonstrate the close relationship between *S. atratum* of *S.* series *Atrata* and *S. brevifolium* of *S.* series *Pedicellata* (van Ham, 1994).

Table 4. Distribution of triterpenes in Eurasian *Sedum*.

	Triterpenes 1)								
	LU	AA	BA	GA	GF	TA	TO	FF	MA
S. ser. *Acria*	●	.	.	.	.	.	.	.	.
S. ser. *Alpestria*	●	.	.	.	.	.	.	.	.
S. ser. *Melananthera*	●	.	.	.	.	.	.	.	.
S. ser. *Macaronesia*	.	●	●	.	.	.	.	.	.
S. ser. *Anglica*	.	.	●	●	.	.	.	.	.
S. ser. *Atrata*									
S. atratum	.	.	.	.	.	.	.	.	●
S. ser. *Pedicellata* 2)									
S. andegavense	.	.	●	.	.	.	.	.	.
S. brevifolium	.	.	.	.	.	.	.	.	●
S. ser. *Dasyphylla*	.	.	●	●	.	.	.	.	.
S. ser. *Rupestria*									
S. amplexicaule	.	.	.	.	.	●	.	●	.
S. forsterianum	.	.	.	.	.	●	.	●	.
S. pruinatum	.	.	.	.	●	.	●	.	.
S. ochroleucum	.	.	.	.	.	.	●	.	.
S. montanum	.	.	.	.	●	●	.	.	.
S. rupestre	.	.	.	.	●	●	.	.	.
S. sediforme	.	.	.	.	●	●	.	.	.

Presence of individual triterpenes is indicated by a dot.

1) LU: lupanes, AA: 67-amyrenyl acetate, BA: ß-amyrenyl acetate, GA: germanicyl acetate, GF: germanicyl formate, TA: taraxeryl acetate, TO: taraxerone, FF: fernenyl formate, MA: multiflorenyl acetate.

2) Waxes of *S. candollei* and *S. pedicellatum* were devoid of triterpenes.

The triterpene variation in *S.* series *Rupestria* led us to examine the interspecific relationships in more detail (Stevens & al., 1994b). The triterpene patterns of the presumed allopolyploid *S. rupestre* ssp. *rupestre* (2n=112) and its putative parents *S. forsterianum* (2n=48) and *S. rupestre* ssp. *erectum* (2n=64) were determined, but failed to provide additional evidence for the hypothesis of a hybrid origin of this taxon ('t Hart & al., 1993), . Both subspecies of *S. rupestre* were very similar with germanicyl formate and taraxeryl acetate being the principal triterpenes, whereas *S. forsterianum* contained fernenyl formate instead of germanicyl formate. The absence of fernane triterpenes in *S. rupestre* ssp. *rupestre* can only be accounted for when *S. rupestre* ssp. *rupestre* originated from a non-glaucous *S. forsterianum* plant and a glaucous *S. rupestre* ssp. *erectum* plant in a single hybridization event. By contrast, artificial hybrids between glaucous plants of *S. forsterianum* and *S. rupestre* ssp. *erectum* combined the triterpene patterns of their parents. Moreover, the artificial hybrids contained an additional fernane triterpene, fernenone, which was lacking in the waxes of *S. forsterianum.* The occurrence of taraxerone in *S. rupestre* ssp. *erectum* led us to suggest that the artificial hybrids inherited an oxidative enzyme system from *S. rupestre* ssp. *erectum* in order to be able to convert taraxerol and fernenol into their corresponding keto compounds.

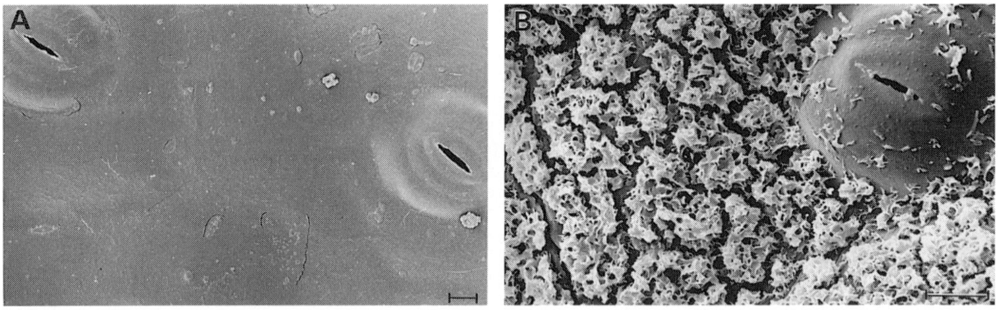

Figure 9. Variation of the ultrastructure of the wax layer in *Sedum rupesture.* 9A. Glossy leaf; wax layer contains less than 1% triterpenes. 9B. Glaucous leaf covered with crystalline platelets; wax layer contains ca 62% triterpenes. Lenght of the bars =10μm

Triterpenoids are virtually absent from epicuticular layers of *Sempervivum* and *Aeonium.* In *Aeonium* triterpenoids have been replaced by diterpenoids, which are formed from geranylgeraniol (C_{20}) and one C_5 unit (triterpenoids are formed through coupling of two C_{15} (C_{10}+C_5) units). The distribution of diterpenes is primarily restricted to the species with sticky leaves, a phenomenon we have also observed in the study of exudate flavonoids of this genus (see above).

The epicuticular diterpenes of *Aeonium* were all identified as derivatives of labdane. Labdane-8,15-diol was previously isolated from the leaf gum of *A. lindleyi* (Baker & al., 1962). We have also isolated this compound from the same species, as well as related labdane isomers from *A. goochiae* (Fig. 11). Using gas chromatography in combination with mass spectrometry, we also detected related diterpenes in epicuticular layers of *A. glutinosum*, *A. spathulatum*, and *A. palmense* from different sections, and all species of *A.* sect. *Petrothamnium* and *A.* sect. *Goochia.* This distribution pattern is very similar to that of the 6-substituted flavonols, which may indicate ancestral inheritance and subsequent loss rather than close relationship. Apparently, the capability of producing (di)terpenoids has been lost in various lineages within *Aeonium.*

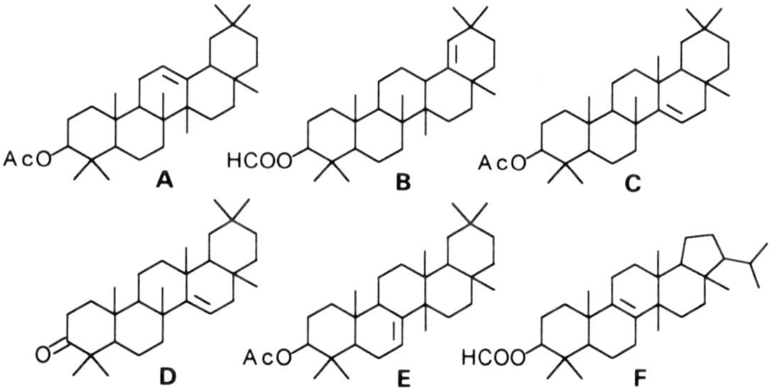

Fig 10. Major triterpenes from waxes of *Sedum.* 10A. β-amyrenyl acetate (oleane series). 10B. Germanicyl formate (oleanane series). 10C. Taraxeryl acetate (taraxerane series). 10D. Taraxerone (taraxerane series). 10E. Multiflorenyl acetate (multiflorane series). 10F. Fernenyl formate (fernane series).

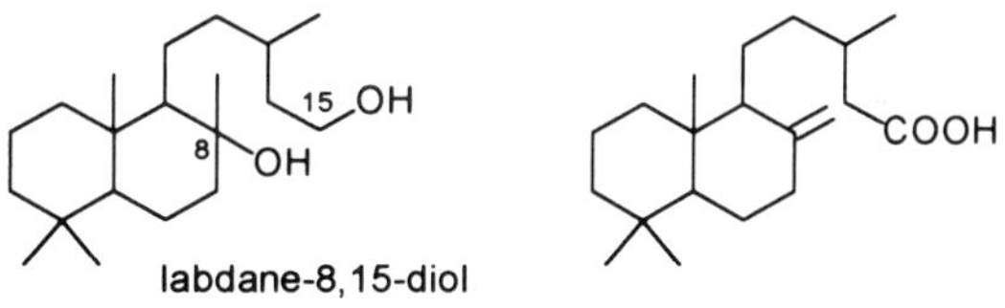

Figure 11. Diterpenes from species of *Aeonium*

Conclusions and final remarks

In the course of our studies, we collected phytochemical data for approximately 120 species of Crassulaceae. To the phytochemical data known from the literature, we added alkaloid, tannin, flavonoid, and terpenoid records (positive and negative) for many species. Most phytochemical papers referred to in Table 1 deal only with the isolation and structure elucidation of a few secondary metabolites. Although these reports were of great help to us, these isolated reports cannot be evaluated for taxonomic purposes due to a combination of incompleteness and differences in methods of analysis, parts of the plants examined, developmental stage of the plants, etc. By contrast, we compared phytochemical patterns of well-documented plants grown under greenhouse conditions (Utrecht Botanic Gardens) and harvested at approximately the same developmental stage (vouchers deposited at the Botanical Institute of the University of Utrecht). Moreover, we had the privilege to compare our phytochemical data with data from morphological, cytological, and molecular studies. This gave us the unique chance to assess the systematic and evolutionary relevance of our chemical data. Our phytochemical approach, despite its largely supportive role, has made a significant contribution to the understanding of the affinities within and between the Eurasian Sedoideae and Sempervivoideae. I will conclude this paper with some highlights of our work and some final remarks.

Most Crassulaceae contain significant amounts of condensed tannins. Within *Sedum*, a switch from the tannins to the production of alkaloids has taken place in species belonging to the *Acre*-clade as distinguished by chloroplast DNA restriction- site variation (van Ham, 1994). Our results convincingly indicate that the alkaloids are restricted to species of this clade, and, in my opinion, any positive alkaloid records for species of Crassulaceae not belonging to this clade should be re-examined. The observed alkaloid/tannin dichotomy has also proved very useful to determine the evolutionary status of some floral and seed characters in *Sedum*. Furthermore, structural variation of the alkaloids is taxonomically significant at the series level. The most remarkable examples are the 2,6-disubstituted piperidines of *S. acre* and the 2-monosubstituted piperidines bearing C_4/C_5 side chains we reported from *S.* series *Macaronesia* for the first time.

The abundant presence of trihydroxylated tannins in most of the Crassulaceae possibly inhibits extensive 'experimentation' with oxygenation in secondary metabolism due to the radical scavenging properties of these tannins. This would explain why the alkaloids and terpenoids of *Sedum* and *Aeonium* originate from condensation reactions rather than from oxidative metabolism. The presence of the trihydroxylated tannins and myricetin in many species emphasizes the incorporation of the Crassulaceae in the Hamamelididae-Rosidae-Dilleniidae complex (Giannasi, 1988). The distinctive phenolic chemistry of this complex led some authors to suggest that Magnoliid-like ancestors may not be primal to the Hamamelididae-Dilleniidae-Rosidae (Giannasi, 1988; Frohne & Jensen, 1992; Gottlieb & al., 1993). Huber (1991) even considered the trihydroxylated phenolics as a gymnospermous heritage.

Flavonoids are useful taxonomic markers at the series level in Sedum and at the section level in *Aeonium*. Within *Sedum*, *S.* series *Rupestria*, *S.* series *Alba*, *S.* series *Atrata*, and part of *S.* series *Alpestria* could be distinguished by multivariate analysis of the flavonoid aglycone data. However, the high degree of parallel evolution in *Sedum*, particularly with respect to 8-hydroxylation and 8-O-methylation, makes any statement as to flavonoid evolution speculative. The exudate flavonoid variation in *Aeonium* enabled us to distinguish *A.* sect. *Aeonium*, *A.* sect *Goochia*, and *A.* sect. *Petrothamnium* from other sections. Rare 6-hydroxyflavonoids are scattered in *Aeonium*, probably as a result of ancestral inheritance followed by loss of 6-hydroxylation in various lineages. Consequently this feature is of little taxonomic significance. *Sempervivum* was found to be extremely uniform in its flavonoid chemistry with kaempferol being the principal flavonol.

The erratic occurrence of epicuticular terpenoids limits their taxonomic applicability, but in *Sedum* triterpenoid variation has great systematic value at the series level. A striking example represents *S.* series *Rupestria*, which could be distinguished from all other infrageneric taxa by the presence of new triterpenyl formates. In *Aeonium* triterpenoids have been replaced by diterpenoids.

Acknowledgements

I am grateful to Dr. H. 't Hart (Utrecht) for supplying the plant samples included in the phytochemical surveys, and for his invaluable support and encouragement during the studies. I also thank Ms. E.T. Elema for technical assistance and Prof. Dr. E. Wollenweber (Darmstadt) for collaboration in the study of exudate flavonoids of *Aeonium*. The investigations were supported by the Foundation for Biological Research (BION), which is subsidized by the Netherlands Organization for Scientific Research (NWO).

References

Baker, A.J., Eglinton, G., Gonzalez, A.G., Hamilton, R.J. & Raphael, R.A. 1962. A new leaf-gum constituent: labdane-8,15-diol. *J. Chem. Soc.*, 4705-4709.

Beyerman, H.C., Eenshuistra, J. & Eveleens, W. 1957. On the absolute configuration of sedamine and sedridine. *Rec. Trav. Chim.* **76**: 415-418.

Bisby, F.A., Vaughan, J.G. & Wright, C.A. (eds.) 1980. Chemosystematics: principles and practice. Academic Press, London.

Bohm, B.A., Bhat, U.G. & Miller, J.M. 1984. Further studies of flavonoids in *Saxifraga. Biochem. Syst. Ecol.* **12**: 367-368.

Borges del Castello, J., Gonzalez, A.G. & Eglinton, G. 1968. Derivados quercetinicos en las ceras de las hojas del *Aeonium manriquorum* Bolle. *An. Fis. y Quim.* **64B**: 193-202.

Bowman, R.N. 1983. Intraspecific variability of leaf cuticle alkanes in *Sedum lanceolatum* along an elevational gradient. *Biochem. Syst. Ecol.* **11**: 195-198.

Calie, P.J. 1981. Systematic studies in *Sedum* section *Ternata* (Crassulaceae). *Brittonia* **33**: 498-507.

Cates, R.G. & Rhoades, D.F. 1977. Patterns in the production of antiherbivore chemical defenses in plant communities. *Biochem. Syst. Ecol.* **5**: 185-193.

Colau, B. & Hootelé, C. 1983. *Sedum* alkaloids. V. Sedacrine, the major alkaloid of *Sedum acre*, and related bases. *Can. J. Chem.* **61**: 470-472.

Coley, P.D., Bryant, J.P. & Chapin, F.S. 1985. Resource availability and plant antiherbivore defence. *Science* **230**: 895-899.

Combier, H., Markham, K., Audier, H., Lebreton, P., Mabry, T. & Jay, M. 1968. Chemotaxonomic research on vascular plants. 3,4',5,7-Tetrahydroxy-8- methoxyflavone extracted from *Sedum acre* var. *sexangulare. Compt. Rend. Acad. Sc., Ser. D* **266**: 2495-2497.

Combier, H. & Jay, M. 1967. Recherches chimiotaxinomiques sur les plantes vasculaire. II. Distribution des flavonoïdes chez les Crassulacées. Plantes médicinales et phytothérapie **1**.

Denton, M.F. & Kerwin, J.L. 1980. Survey of vegetative flavonoids of *Sedum* section *Gormania* (Crassulaceae). *Can. J. Bot.* **58**: 902-905.

Diak, J. & Kohlmünzer, S. 1981. Search for alkaloids in the genus *Sedum* L. *Herba Hungarica* **20**: 7-13.

Driessens, F. & Hootelé, C. 1991. *Sedum* alkaloids. XI. Synthesis of sedinone and sedacrine by application of anodic oxidation. *Can. J. Chem.* **69**: 211-217.

Eglinton, G., Gonzalez, A.G., Hamilton, R.J. & Raphael, R.A. 1962. Hydrocarbon constituents of the wax coatings of plant leaves: a taxonomic survey. *Phytochemistry* **1**: 89-102.

Ehrlich, P.R. & Raven, P.H. 1965. Butterflies and plants: a study in coevolution. *Evolution* **18**: 586-608.

Fang Shen-diu, Yan Xin-quan, Li Ching-fang, Fan Zhi-yun, Xy Xia-yi, & Xu Jen-sheng 1979. *Kexue Tongbao* **24**: 431.

Francis, L.P.S. & Francis, G.W. 1977. *Sedum* alkaloids. 1. Polar alkaloids in *Sedum acre* L. *Planta Medica* **32**: 268-274.

Franck, B. 1958. Alkaloide in *Sedum acre* und verwandten Sedumarten. *Chem. Ber.* **91**: 2803-2818.

Frigot, P. 1960. Contribution à l'etude chimique de quelques Crassulacées et en particulier de divers *Sedum.* Thesis, Paris.

Frohne, D. & Jensen, U. 1992. Systematik des Pflanzenreichs, ed. 4. Gustav Fischer, Stuttgart.

Giannasi, D.E. 1988. Flavonoids and evolution in the dicotyledons. In Harborne, J.B. (ed.), The flavonoids, 479-504. Chapman & Hall, London.

Giannasi, D.E. & Niklas, K.J. 1977. Flavonoid and other chemical constituents of fossil Miocene Celtis and Ulmus (Succor Creek Flora). *Science* **197**: 765-767.

Gill, S., Raszeja, W. & Szynkiewicz, G. 1979. Occurrence of nicotine in some species of the genus *Sedum. Farm. Pol.* **35**: 151-153.

Gottlieb, O.R. 1982a. Micromolecular evolution, systematics and ecology. Springer-Verlag, Berlin.

Gottlieb, O.R. 1982b. Ethnopharmacology versus chemosystematics in the search for biologically active principles in plants. *J. Ethnopharmacology* **6**: 227-238.

Gottlieb, O.R., Kaplan, M.A.C. & Kubitzki, K. 1993. A suggested role of galloyl esters in the evolution of dicotyledons. *Taxon* **42**: 539-552.

Grayer, R.J., Kimmins, F.M., Padgham, D.E., Harborne, J.B. & Rao, D.V.R. 1992. Condensed tannin levels and resistance of groundnuts (*Arachis hypogaea*) against *Aphis craccivora. Phytochemistry* **31**: 3795-3800.

Gumenyuk, L.A., Gnedkov, P.A. & Batyuk, V.S. 1970. Kaempferol and quercetin from *Sempervivum ruthenicum. Khim. Prir. Soedin.* **6**: 630.

Gumenyuk, L.A., Gnedkov, P.A. & Batyuk, V.S. 1971. Astragalin from *Sempervivum ruthenicum. Khim. Prir. Soedin.* **7**: 202.

Gumenyuk, L.A., Dykhanov, N.N. & Batyuk, V.S. 1972. Flavonoid compounds from blossoms of *Sempervivum ruthenicum. Khim. Prir. Soedin.* **8**: 391-392.

Gumenyuk, L.A., Golovataya, V.F., Perova, T.V. & Bagrii, A.K. 1982. Phytochemical study on *Sempervivum ruthenicum. Farm. Zh.*, 61-65.

Gupta, R.N. & Spenser, I.D. 1967. The biosynthesis of sedamine. *Can. J. Chem.* **45**: 1275-1285.

Gupta, R.N. & Spenser, I.D. 1969. Biosynthesis of N-methylpelletierine. *Phytochemistry* **8**: 1937-1944.

Halin, F., Slosse, P. & Hootelé, C. 1985. *Sedum* alkaloids. VII. Structure and synthesis of (+)-4-hydroxysedamine and (+)-4-hydroxyallosedamine. *Tetrahedron* **41**: 2891-2897.

Ham, R.C.H.J. van 1994. Phylogenetic implications of chloroplast DNA variation in the Crassulaceae. Ph.D. Thesis, Utrecht University.

Harborne, J.B. 1988. Introduction to Ecological Biochemistry, ed. 3. Academic Press, London.

Harborne, J.B. & Turner, B.L. 1984. Plant chemosystematics. Academic Press, London.

Harborne, J.B. & Baxter, H. (eds.) 1993. Phytochemical dictionary. Taylor & Francis, London.

Hart, H. 't. 1991. Evolution and classification of the European *Sedum* species (Crassulaceae). *Flora Medit.* **1**: 31-61.

Hart, H. 't. & Koek-Noorman, J. 1989. The origin of the woody Sedoideae (Crassulaceae). *Taxon* **38**: 535-544.

Hart, H. 't, Sandbrink, J.M., Csikos, I., Ooyen, A. van, & Brederode, J. van. 1993. The allopolyploid origin of *Sedum rupestre* subsp. *rupestre* (Crassulaceae). *Pl. Syst. Evol.* **184**: 195-206.
Hart, H. 't. The evolution of the *Sedum acre*-group (Crassulaceae). *Flora Medit.* (in press).
Haslam, E. 1982. Proanthocyanidins. In Harborne, J.B. & Mabry, T.J. (eds.), The flavonoids. Advances in research, 417-448. Chapman & Hall, London.
Haslam, E. 1989. Plant polyphenols. Vegetable tannins revisited. Cambridge University Press, Cambridge.
Hegnauer, R. 1962-1994. Chemotaxonomie der Pflanzen, **1-11**. Birkhäuser, Basel.
Hegnauer, R. 1986. Phytochemistry and plant taxonomy. An essay on the chemotaxonomy of higher plants (review). *Phytochemistry* **25**: 1519-1535.
Heller, W. & Forkmann, G. 1994. Biosynthesis of flavonoids. In Harborne, J.B. (ed.), The flavonoids: advances in research since 1986, 499-535. Chapman & Hall, London.
Herbin, G.A. & Robins, P.A. 1969. Patterns of variation and development in leaf wax alkanes. *Phytochemistry* **8**: 1985-1998.
Hootelé, C., Etienne, J.P. & Colau, B. 1979. Sederine, a new alkaloid from *Sedum acre*. *Bull. Soc. Chim. Belg.* **88**: 111-112.
Hootelé, C., Colau, B., Halin, F., Declercq, J.P., Germain, G. & Van Meerssche, M. 1980a. *Sedum* alkaloids. 11. Sedacryptine, a new minor base from *Sedum acre*. *Tetrahedron Letters* **21**: 5061-5062.
Hootelé, C., Colau, B., Halin, F., Declercq, J.P., Germain, G. & Van Meerssche, M. 1980b. *Sedum* alkaloids. III. A revised structure for sedinine. *Tetrahedron Letters* **21**: 5063-5066.
Huber, H. 1991. Angiospermen. Gustav Fischer, Stuttgart.
Humphries, C.J. & Richardson, P.M. 1980. Hennig's methods and phytochemistry. In Bisby, F.A., Vaughan, J.G. & Wright, C.A. (eds.), Chemosystematics: principles and practice, 353-378. Academic Press, London.
Ibebeke-Bomangwa, W. & Hootelé, C. 1987. *Sedum* alkaloids. X. Structure and synthesis of new 3- and 5-hydroxypiperidine alkaloids. *Tetrahedron* **43**: 935-945.
Jacobs, M. & Rubery, P.H. 1988. Naturally occurring auxin transport regulators. *Science* **241**: 346-349.
Jay, M. 1970. Quelques problèmes taxinomiques et phylogénétiques des Saxifragacées vus à la lumière de la biochimie flavonique. *Bull. Mus. Nation. Hist. Natur.* **42**: 754-775.
Khetwal, K.S., Manral, K. & Verma, D.L. 1988. Flavonoids from the roots of *Sedum crassipes*. *Planta Medica* **54**: 89-90.
Kooy J.H. 1976. Alkaloids of *Sedum acre*. *Planta Medica* **30**: 295-296.
Krasnov, E.A., Petrova, L.V. & Bekker, E.F. 1977. Alkaloids of *Sedum* species. *Khim. Prir. Soedin.* **4**: 585.
Krolikowska, M. 1966. The investigation of the flavonoid constituents of the herb *Sedum acre* L. (Crassulaceae). *Roczn. Chem.* **40**: 717-719.
Krolikowska, M. 1972. New flavonoid compounds from the herb *Sedum acre* L. (Crassulaceae). *Roczn. Chem.* **46**: 969-970.
Lebreton, P. 1982. Tanins ou alcaloides: deux tactiques phytochimiques de dissuasion des herbivores. *Rev. Ecol. (Terre Vie)* **36**: 539-572.
Liu, H.Y. 1989. Systematics of *Aeonium* (Crassulaceae). Special publications number **3**, National Museum of Natural Science, Taiwan.
Logar, S., Mesicek, N., Perpar, M. & Seles, E. 1974. Alkaloids in the orpine (*Sedum maximum*). *Farm. Vestn. (Ljubljana)* **25**: 21-25.
Maksimovic, M., Sober, M. & Nikolin, B. 1990. Gas chromatography/mass spectrometry in the elucidation of the structure of piperidine alkaloids. *Rapid Commun. Mass Spectrom.* **4**: 503-504.
Malterud, K.E. & Nordal, A. 1991. Structure elucidation of sedoflorigenin, a flavonoid from *Sedum*

acre. Acta Pharm. Nord. **3**: 99-100.

Manheim, B.S. & Mulroy, T.W. 1978. Triterpenoids in epicuticular waxes of *Dudleya* species. *Phytochemistry* **17**: 1799-1800.

Mann, J. 1987. Secondary metabolism, ed. 2. Clarendon Press, Oxford.

Mann, J. 1994. Murder, Magic, and Medicine. Oxford University Press, Oxford.

Marion, L. 1945. The alkaloids of *Sedum acre* L. *Can. J. Res.* **23B**: 165-166.

Marion, L. & Chaput, M. 1949. A new occurrence of dl-methylisopelletierine. *Can. J. Res.* **27B**: 215-217.

Marion, L., Lavigne, R. & Lemay, L. 1951. The structure of sedamine. *Can. J. Chem.* **29**: 347-351.

Martin, J.T. & Juniper, B.E. 1970. The Cuticles of Plants. Edward Arnold, London.

McClure, J.W. 1975. Physiology and functions of flavonoids. In Harborne, J.B., Mabry, T.J. & Mabry, H. (eds.), The flavonoids, 970-1055. Chapman & Hall, London.

Mnajed, H., Raynaud, J. & Combier, H. 1972. New natural heteroside of 8-methoxykaempferol extracted from *Sedum sexangulare* (Crassulaceae). *Compt. Rend. Acad. Sci., Ser. D* **274**: 445-448.

Nahrstedt, A., Walther, A., Wray, V. 1982. Sarmentosin epoxide, a new cyanogenic compound from *Sedum cepaea. Phytochemistry* **21**: 107-110.

Niemann, G.J., Visser-Simons, J.M.J. & Hart, H. 't. 1976. Flavonoids of some species of *Sedum. Planta Medica* **30**: 384-387.

Nordal, A. 1946. A pharmacognostical study of *Sedum acre* L. Johan Grundt Tanum, Oslo.

Piccinni-Leopardi, C., Tinant, B., Declercq, J.P., Van Meerssche, M. & Hootelé, C. 1987. *Sedum* alkaloids. IX. Isolation from *Sedum acre* and X-ray structure determination of lobelanidine glycoside. *Bull. Soc. Chim. Belg.* **96**: 97-100.

Porter, L.J. 1988. Flavans and proanthocyanidins. In Harborne, J.B. (ed.), The flavonoids: advances in research since 1980, 21-62. Chapman & Hall, London.

Porter, L.J. 1989. Tannins. In Harborne, J.B. (ed.), Methods in plant biochemistry, **1**. Plant phenolics, 389-419. Academic Press, London.

Porter, L.J., Hrstich, L.N. & Chan, B.G. 1986. The conversion of procyanidins and prodelphinidins to cyanidin and delphinidin. *Phytochemistry* **25**: 223-230.

Richardson, P.M. & Young, D.A. 1982. The phylogenetic content of flavonoid point scores. *Biochem. Syst. Ecol.* **10**: 251-255.

Rodriguez, E., Healy, P.L. & Mehta, I. 1984. Biology and chemistry of plant trichomes. Plenum, New York.

Sakar, M.K., Petereit, F. & Nahrstedt, A. 1993. Two phloroglucinol glucosides, flavan gallates and flavonol glycosides from *Sedum sediforme* flowers. *Phytochemistry* **33**: 171-174.

Schöpf, C. & Unger, R. 1956. Über physiologische, durch einen Gehalt an verschiedenen Alkaloiden charakterisierte Rassen von *Sedum acre* L. *Experientia* **12**: 19-20.

Schöpf, C. & Unger, R. 1957. Thesis, Techn. Hochschule Darmstadt.

Shnyakina, G.P. & Zapesochnaja, G.G. 1973a. *Khim. Prir. Soedin.* **9**: 557.

Shnyakina, G.P. & Zapesochnaja, G.G. 1973b. *Khim. Prir. Soedin.* **9**: 674.

Stevens, J.F., Hart, H. 't, Hendriks, H. & Malingré, T.M. 1992. Alkaloids of some European and Macaronesian Sedoideae and Sempervivoideae (Crassulaceae). Phytochemistry **31**: 3917-3924.

Stevens, J.F., Hart, H. 't, Hendriks, H. & Malingré, T.M. 1993. Alkaloids of the *Sedum acre*-group (Crassulaceae). *Pl. Syst. Evol.* **185**: 207-217.

Stevens, J.F., Hart, H. 't, Bolck, A., Zwaving, J.H. & Malingré, T.M. 1994a. Epicuticular wax composition of some European *Sedum* species. *Phytochemistry* **35**: 389-399.

Stevens, J.F., Hart, H. 't, Pouw, A.J.A., Bolck, A. & Zwaving, J.H. 1994b. Epicuticular waxes of *Sedum* series *Rupestria. Phytochemistry* **36**: 341-348.

Stevens, J.F., Elema, E.T. & Hart, H. 't. 1994c. Myricetin 3-O-arabinofuranoside from *Sedum montanum* ssp. *orientale. Biochem. Syst. Ecol.* **22**: 861-862.

Stevens, J.F., Hart, H. 't, Ham, R.C.H.J. van, Elema, E.T., Ent, M.M.V.X. van den, Wildeboer, M. & Zwaving, J.H. 1995a. Distribution of alkaloids and tannins in the Crassulaceae. *Biochem. Syst. Ecol.* **23**: 157-165.

Stevens, J.F., Hart, H.'t & Wollenweber, E. 1995b. The systematic and evolutionary significance of exudate flavonoids in *Aeonium*. *Phytochemistry*. **39**: 805-813.

Stuessy, T.F. 1990. Plant taxonomy. Columbia University Press, New York.

Stuessy, T.F & Crawford, D.J. 1983. Flavonoids and phylogenetic reconstruction. *Pl. Syst. Evol.* **143**: 83-107.

Wal, R. van der, Kooy, J.H. & Eijk, J.L. van. 1981. Phytochemical investigation of *Sedum acre* L. *Planta Medica* **43**: 97-99.

Waterman, P.G. & Gray, A.I. 1987. Chemical systematics. *Nat. Prod. Rep.* **4**: 175-203.

Whittaker, R.H. & Feeny, P.P. 1971. Allelochemics: chemical interactions between species. *Science* **171**: 757-770.

Williams, C.A. 1978. The systematic implications of the complexity of leaf flavonoids in the Bromeliaceae. *Phytochemistry* **17**: 729-734.

Williams, C.A., Harborne, J.B. & Goldblatt, P. 1986. Correlations between phenolic patterns and tribal classification in the family Iridaceae. *Phytochemistry* **25**: 2135-2154.

Williams, C.A. & Harborne, J.B. 1994. Flavone and flavonol glycosides. In Harborne, J.B. (ed.), The flavonoids: advances in research since 1986, 337-385. Chapman & Hall, London.

Wolbis, M. 1989. Flavonol glycosides from *Sedum album*. *Phytochemistry* **28**: 2187-2189.

Wolbis, M. & Krolikowska, M. 1988. Flavonol glycosides from *Sedum acre*. *Phytochemistry* **27**: 3941-3943.

Wolbis, M. & Kawalec, M. 1990. Investigation of phenolic compounds in *Sedum sexangulare* (Crassulaceae). *Acta Pol. Pharm.* **47**: 81-82.

Wollenweber, E. 1990. On the distribution of exudate flavonoids among angiosperms. *Rev. Latinoam. Quim.* **21**: 115-121.

Wollenweber, E. 1994. Flavones and flavonols. In Harborne, J.B. (ed.), The flavonoids: advances in research since 1986, 259-335. Chapman & Hall, London.

CHAPTER 5

Hybridization in *Monanthes*

Reto Nyffeler

Abstract

Artificial crossing and backcrossing experiments involving 8 of the 9 perennial taxa of *Monanthes* (Crassulaceae) revealed that the members of this genus form a comparium. Most taxa can be easily crossed and seedset is not significantly lower than in infraspecific crosses. The F1 hybrids are diploid, triploid or tetraploid, dependent on the ploidy level of the parents. Backcrosses of the diploid F1 hybrids with both parents were successful, though the seedlings died in an early stage. The study of pollen fertility with viability staining proved to be of considerable value to evaluate the probable hybrid origin of a specimen. Most artificial hybrids showed reduced fertility, though pollen fertility ranged from completely sterile to highly fertile. The percentage of stainable pollen was larger in diploid than in tetraploid F1 hybrids, and was less than 50% in triploid hybrids. Triploid hybrids produced some exceptionally large pollen grains which are probably unreduced gametes. Hybridization in nature is discussed for all combinations of taxa known to occur together as well as for all other combinations reported to produce hybrids. Plants of unquestionable hybrid origin from Tenerife can not always be simply explained as F1 hybrids when compared with the artificial hybrids. Introgression is thought to be the cause of the often complex patterns of morphological variation, and the intermediate individuals found at locations where two or three taxa occur together. Finally, a short digression on the naming of hybrids is given in which the use of hybrid formulae in *Monanthes* is strongly recommended.

Introduction

The genus *Monanthes* comprises 9 species and 1 subspecies (Nyffeler, 1992) as recent molecular studies showed that *M. atlantica* Ball. (= *Sedum surculosum* Coss.) should be excluded from *Monanthes* (Mes & 't Hart, 1994). The genus is endemic to the Canary Islands and the Salvage Islands. As many other endemic groups of the Macaronesian phytogeographical region, the members of *Monanthes* underwent a diversification into a wide range of growth-forms.

Extensive interspecific hybridization is known to occur among the taxa of the different genera of Macaronesian Sempervivoideae, i.e. *Aeonium* (including *Greenovia*), *Aichryson*, and *Monanthes*. The full extent of this phenomenon was first recognized by Burchard (1929) and Praeger (1929). Since that time many morphologically intermediate plants have been interpreted as interspecific hybrids (Praeger, 1929; Bañares, 1990) and have often been given binominal names (Jacobsen & Rowley, 1973; Bañares, 1990; Heath 1994).

Although Praeger (1929) already emphasized the importance of artificial crosses and backcrosses (if possible) to confirm the postulated hybrids, no such studies have been carried out. The motives for the present hybridization experiments have been on the one hand simply to give experimental proof for hybridization in *Monanthes*, and on the other hand to learn more about the mechanisms involved in the origin of the sometimes complex patterns of morphological variation, and the intermediate plants observed at several localities in nature.

Material and methods

Eight perennial taxa of *Monanthes* were included in the crossing experiment. *M. minima* and the annual *M. icterica* were not available at the time. Emphasis was given to crosses between taxa which were found growing together by the author in nature as well as to taxa which have been reported to produce natural hybrids. The chromosome numbers of the plants used in these experiments were not known in advance. This was not considered disadvantageous as there is little cytological variation in *Monanthes* (Uhl, 1961b). In a few cases the experiments were hampered by differences in flowering periods (especially in experiments including *M. anagensis*) or by the small number of flowers produced by some taxa.

All plants used in this study were collected in nature. The collection data are given in Table 1. The plants were cultivated under uniform conditions in the experimental garden of the University of Zürich. Voucher specimens are deposited at the Institute for Systematic Botany, University of Zürich (Z). The crossing experiments were carried out in 1991 and 1993. Chromosome numbers were kindly determined by Dr. H. 't Hart, University of Utrecht. Pollen fertility was determined by staining fresh pollen samples with aceto-carmine in glycerine (Radford & al. 1974). Of each hybrid at least two flowers were used.

Table 1. Chromosome number and origin of the plants used in the hybridization experiments

Monanthes anagensis Praeger, 2n=108
Tenerife; Anaga, E of Las Carboneras, Lösch *s.n.*, s.d.
Monanthes brachycaulos (Webb & Berthelot) Lowe, 2n=36
Gran Canaria; between Firgas and Moya, Kuschel *90-2*, 1990.
Tenerife; Las Cañadas, Rocas de la Fortaleza, 2 km NW of Portillo, 2075 m, Nyffeler *143*, 1990.
Tenerife; Aguamansa, 1120 m, Thiede *2034*, 1989.
Monanthes laxiflora (De Candolle) Bornmüller, 2n=36
Tenerife; Anaga region, El Bailadero, 750 m, Nyffeler *122*, 1990.
Monanthes lowei (Paiva) Pérez & Acebes, 2n=36
Selvagens Islands, Anonymus *s.n.*, s.d. (ex cult. Hort. Bot. Hamburg).
Monanthes muralis (Bolle) Hooker fil., 2n=72
Hierro; descent to El Golfo, on cliffs along footpath, 900 m, Kuschel *90-5*, 1990.
Monanthes pallens (Webb) Christ, 2n=36
Tenerife; Barranco del Infierno (ca. 2 km above the entry), 600 m, Thiede *2089*, 1989.
Monanthes polyphylla Haworth ssp. *amydros* Nyffeler, 2n=36
Gomera; Degollada de la Cumbre, rocks NW of the N portal of tunnel, 630 m, Nyffeler *152*, 1990.
Monanthes polyphylla Haworth ssp. *polyphylla*, 2n=72
Tenerife; Teno region, 500 m W of the long tunnel to Punta de Teno, 160 m, Thiede *2019*, 1989.

Results and discussion

Crossing experiments - In contrast to many other groups of Crassulaceae *Monanthes* is cytologically rather uniform (Uhl, 1961a). All perennial taxa have a basic chromosome number of x=18 (Uhl, 1961b; 't Hart & Nyffeler, unpublished data). Cytologically the annual *M. icterica* is quite different, it has the somatic chromosome number is 2n=20 ('t Hart & Nyffeler, unpublished data). In the perennial *Monanthes* taxa occur the chromosome numbers 2n=36, 72, and 108 (Tab. 1). More than one cytotype has been reported for *M. anagensis* and *M. polyphylla* ssp. *polyphylla* (Uhl, 1961b; 't Hart & Nyffeler, unpublished data). Dysploidy is not known to occur. The perennial taxa of *M.* sect. *Petrophyllae* (*M. brachycaulos*, *M. lowei*, *M. minima*, and *M. pallens*) are all diploid (2n=36). *M.* sect. *Monanthes* and *M.* sect. *Sedoideae*, on the other hand, contain diploids (*M. polyphylla* ssp. *amydros* and *M. polyphylla* ssp. *polyphylla* of *M.* sect. *Monanthes*; *M. laxiflora* of *M.* sect. *Sedoideae*) as well as tetraploid (*M. muralis* and *M. polyphylla* ssp. *polyphylla* of *M.* sect. *Monanthes*; *M. anagensis* of *M.* sect. *Sedoideae*) and hexaploid taxa (*M. anagensis*).

The results of the crossing experiments are summarized in Figure 1. About half of the crosses produced viable seeds which gave rise to vigorous F1 plants, though *M. anagensis* x *M. laxiflora* died before flowering. Only limited data was recorded on the percentages of germinated seeds. However, in general the percentage of germinating seeds proved to be considerably lower than for seeds obtained from infraspecific crosses in which often almost 100% germination is observed. Only the reciprocal crosses with *M. muralis* and *M. polyphylla* ssp. *polyphylla* were both successful, in all other reciprocal crosses one of the combinations failed.

f m	AMY 2n = 36	ANA 2n = 108	BRA 2n = 36	LAX 2n = 36	MUR 2n = 72	PAL 2n = 36	POL 2n = 72
AMY 2n = 36				2/1/6/0			
ANA 2n = 108				1/0/0/0			
BRA 2n = 36	1/0/0/0			**1/1/19/11** 2n = 36	**2/1/?/?** 2n = 54	**1/1/25/?** 2n = 36	
LAX 2n = 36	1/0/0/0	**1/1/47/?**	1/0/0/0				1/0/0/0
LOW 2n = 36			2/1/31/0	1/1/5/0		1/1/6/0	1/1/1/0
MUR 2n = 72	**1/1/?/?** 2n = 54			**1/1/19/?** 2n = 54			**2/1/?/?** 2n = 72
PAL 2n = 36				**1/1/16/?** 2n = 36			
POL 2n = 72	**1/1/?/?** 2n = 54		**1/1/?/?** 2n = 54	**1/1/12/?** 2n = 54	**1/1/36/?** 2n = 72	**3/2/16/10** 2n = 54	

Figure 1. Results of crossing experiments with *Monanthes polyphylla* ssp. *amydros* (AMY), *M. anagensis* (ANA), *M. brachycaulos* (BRA), *M. laxiflora* (LAX), *M. lowei* (LOW), *M. muralis* (MUR), *M. pallens* (PAL), and *M. polyphylla* ssp. *polyphylla* (POL). The first figure in each square indicates the number of flowers used in each cross, the second the number of flowers producing seeds, the third the total number of seeds produced, and the fourth the number of germinated seeds. The successful crossings are printed in bold face. Somatic chromosome numbers of the parents and F1 hybrids are indicated.

f m	BRA × LAX 2n = 36	BRA × POL 2n = 54	LAX × PAL 2n = 36	LAX × POL 2n = 54	PAL × POL 2n = 54
BRA 2n = 36	**4/3/42/15**				
LAX 2n = 36	**4/4/41/24**	2/0/0/0	**2/2/35/17**	1/1/10/0	2/1/1/0
PAL 2n = 36			**3/3/33/5**	1/0/0/0	
POL 2n = 72	1/0/0/0	1/0/0/0	1/0/0/0	1/0/0/0	
BRA × POL 2n = 54	1/0/0/0				
LAX × POL 2n = 54			3/0/0/0		

Figure 2. Results of backcrossing experiments in *Monanthes* (see Figure 1 for acronyms of taxa and meaning of the figures in each square).

Some of the F1 hybrids were used as parents in backcrossing experiments. The pollen of the hybrids was in most cases visibly poor and difficult to handle. Therefore the F1 hybrids were used as female parents. The results of the backcrossing experiments are given in Figure 2. Only backcrosses between diploid hybrids and their respective parents produced viable offspring. Seedset in these experiments did not differ significantly from crosses among the parental plants and the seeds were all of normal shape and size. Up to 50% of the seeds germinated, but, unfortunately, the seedlings died in an early stage. Reduced viability of the seedlings and insufficient care may be the reasons for the demise of the plantlets. Backcrossing of triploid F1 hybrids with *M. laxiflora* produced a few seeds in two combinations, but these did not germinate.

In view of the low number of cross-pollinations carried out for each combination the unsuccessful crosses are necessarily only of very little value and it is expected that eventually the unsuccessful combinations will also yield seeds if the experiments are continued. From the present results it already becomes obvious that the perennial taxa of *Monanthes* (despite the negative results for *M. lowei* and the exclusion of *M. minima* from the experiment) have the potential to hybridize with one another as the crossing polygon in Figure 3 clearly illustrates. The perennial taxa of *Monanthes* form a comparium or syngamodeme (a group of taxa capable to hybridize with one another). The Crassulaceae are well known for their widely retained crossability, which is most impressively exemplified by the large Mexican comparium including over one hundred species of *Echeveria, Graptopetalum, Pachyphytum, Sedum,* and *Villadia* (Uhl, 1992), and *Monanthes* is no exception. In this study the boundaries of the *Monanthes* comparium have not yet been determined. The annual *M. icterica* is of special interest here.

Hybrid fertility - Viable pollen is fundamental to the reproductive process and therefore assessment of pollen fertility is essential in hybridization experiments. Staining of the pollen grains is a simple method to get a rough estimate of their viability as the dye is absorbed to specific cell constituents only present in viable pollen grains (Stanley & Linskens, 1974). The fertility of hybrids is generally impaired and the percentage of stainable pollen is a good indication of their fertility. In *Monanthes* hybrid pollen fertility varies widely, from completely sterile to fully fertile. Pollen fertility of the hybrids is mainly affected by the levels of ploidy of the parental plants and their genetic (systematic) relationship. Furthermore, the quality of the pollen may be affected by the condition of the plants, and also varies among individuals of the same progeny.

The morphology of the pollen grains may also indicate the hybrid nature of a specimen. The pollen grains of a fully fertile plant are usually all stained, of equal shape and size, and regularly distributed on a microscopic slide (Fig. 4A). Hybrids, on the other hand, usually produce variable numbers of unstained (empty), often considerably smaller grains which have no living contents (Fig. 4B). Triploid hybrids, moreover, often produce some extra large, irregularly shaped pollen with four or sometimes more colpi (Fig. 4C). The pollen of some hybrids is apparently completely non-functional, it is sticky, forming dense clusters of faintly stained, irregular, small grains (Fig. 4D).

The results of the pollen fertility studies are summarized in Table 2. Three classes of pollen fertilities are distinguished which are based on stainability and morphology of the pollen grains. 1) High: stainability more than 70%, pollen grains regularly shaped. 2) Medium: stainability 40-70%, pollen grains regularly shaped or small (rarely with extra large grains). 3) Low: stainability less than 40%, pollen grains often sticky, samples always with extra large grains. In the crossing polygon of Figure 3 the fertility of the hybrids is indicated using these three classes.

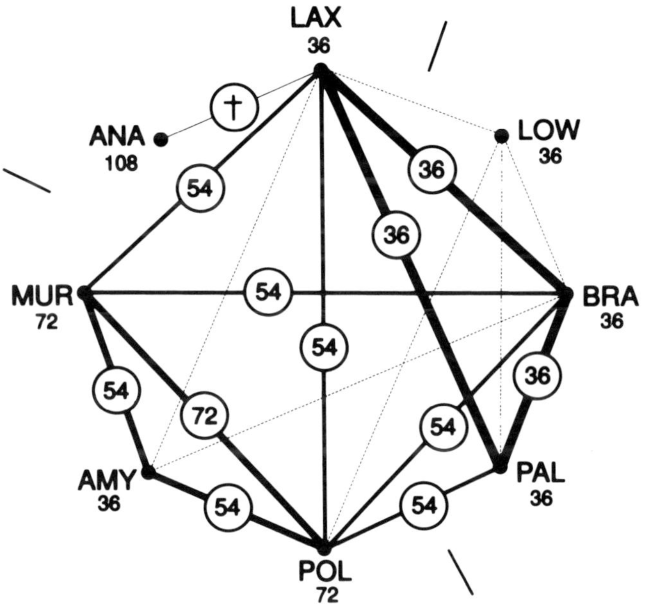

Figure 3. Crossing polygon of eight perennial taxa of *Monanthes* (see Figure 1 for acronyms of the taxa). Figures indicate somatic chromosome numbers. Dotted lines indicate the unsuccessful crosses (no viable offpring). Pollen feritlity of the successful crosses is indicated by increasing boldness of the solid lines: 1) not determined; 2) low (≤ 40%); 3) medium (40 - 70%); 4) high (≥ 70%).

The three diploid hybrids *M. brachycaulos* x *laxiflora*, *M. brachycaulos* x *pallens*, and *M. laxiflora* x *pallens* are highly fertile (Table 2). In contrast to the diploid hybrids, the fertility of the tetraploid hybrid *M. muralis* x *polyphylla* ssp. *polyphylla* is surprisingly low. The extra large pollen grains found in most triploid hybrids are most probably unreduced (with two or three chromosome sets) or aneuploid gametes resulting from imperfect meiosis. The fertility of the triploid hybrids is considerably impaired (pollen stainability 22-54%). However, fertility of triploid intersectional hybrids tends to be somewhat lower than of hybrids between taxa of the same section. If the pollen fertility of the F1 hybrids is actually inversely related to the genetic distance between their parents, the results support Nyffeler's (1992) delimitation of *M.* sect. *Monanthes*. Studies in *Sedum* by 't Hart (1987) showed that spontaneous amphiploids occur among the offspring of hybrids with odd ploidy levels. That this also occurs in *Monanthes* still needs to be confirmed.

The study of pollen fertility proved to be of great value in tracing the presumed hybrid origin of a specimen. Additionally, the percentage of viable pollen grains and pollen morphology give some further evidence, allowing careful statements on the nature of the hybrids. A study of the pollen morphology is also of considerable help for the identification of herbarium specimens, when hybridization is thought to be involved.

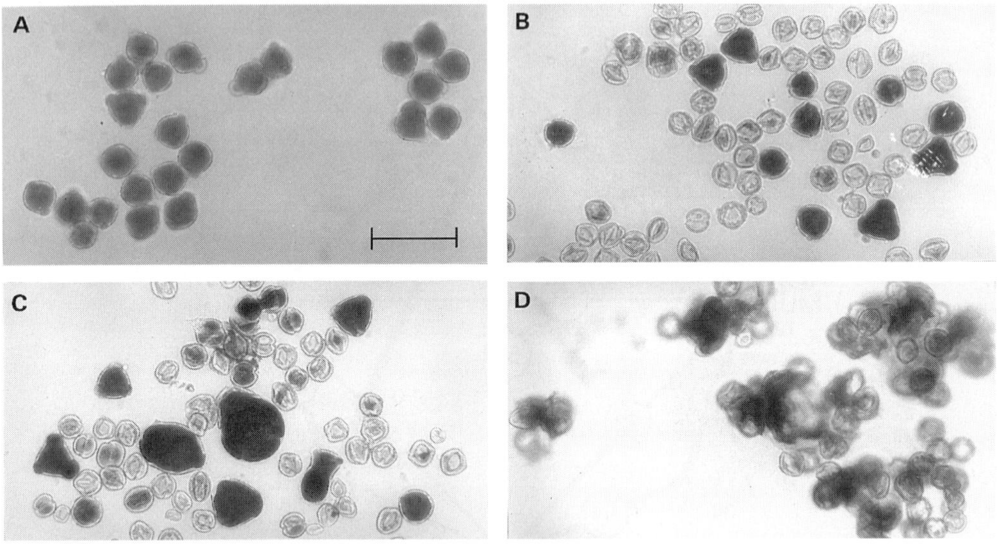

Figure 4. Examples of pollen fertility in *Monanthes* (hybrids). 4a. Fully fertile pollen from *M. Muralis* (Kuschel 90-5). 4B. Sample with low pollen fertility from *M. brachycaulos* x *muralis*. 4C. Sample with unreduced or partly unreduced pollen from triploid *M. laxiflora* x *polyphylla* ssp. *polyphylla*. 4D. Clumps of sticky, imperfecy pollen in a largeley sterile sample from *M. pallens* x *polyphylla* ssp. *polyphylla*. Lenght of the bar= 0.05 mm.

Hybrid morphology - Morphologically the diploid artificial F1 hybrids are intermediate between their parents. This is most apparent in the phyllotaxis of leaf rosettes. Additionally, leaf shape and distribution of bladder-cell idioblasts may be indicative. When the parents belong to different sections, the recognition of hybrids poses no problems to those familiar with *Monanthes*, e.g. all hybrids involving *M. laxiflora* can be recognized in most cases at a glance. However, when both parents belong to the same section the identification of hybrids is more difficult, e.g. *M. brachycaulos* x *pallens*. In these cases the study of pollen fertility could be very helpful.

The triploid F1 hybrids resemble their tetraploid parents most closely. This may not come as a surprise, because tetraploid parents contribute twice the number of chromosomes of the diploid parents. To recognize the hybrid nature of these plants is therefore much more difficult, and in these cases a study of pollen fertility is of great value. Especially when the parental species are very close morphologically, e.g. *M. pallens* and *M. polyphylla* ssp. *polyphylla* which have very similar phyllotaxis. The inflorescences of

Table 2. Pollen fertility of F1 *Monanthes* hybrids. Chromosome number, percentage of stained pollen, and presence of extra large and sticky pollen [+ present; (+) partly present; - absent]. See Figure 1 for acronyms of the taxa.

Hybrid	2n=	stained	extra large	sticky
AMY x MUR	54	54	(+)	+
AMY x POL	54	53	(+)	+
BRA x LAX	36	74	-	-
BRA x MUR	54	35	+	(+)
BRA x PAL	36	98	-	-
BRA x POL	54	32	+	+
LAX x MUR	54	45	(+)	+
LAX x PAL	36	82	-	-
LAX x POL	54	34	+	+
MUR x POL	72	64	-	-
PAL x POL	54	22	-	-

the triploid hybrids between (diploid) taxa of *M.* sect. *Petrophyllae* (*M. brachycaulos* and *M. pallens*) and *M. polyphylla* ssp. *polyphylla* (tetraploid) always arise from the centre of the rosettes similar to the tetraploid parent. In some cases *M. brachycaulos* x *polyphylla* ssp. *polyphylla* produced a terminal inflorescence on a rosette of obovate leaves similar to the leaves on the floriferous shoots of *M. brachycaulos.*

Morphologically the artificial hybrids proved to be very plastic. The expression of certain features largely depends on their condition. Comparison of hybrids obtained by artificial crossings with plants found in nature is therefore somewhat problematic. Cultivation experiments are needed to eliminate this plasticity in comparative studies.

Natural hybrids - In nature hybrids are found at most localities with two or more sympatric taxa. They are restricted to the common area of their parents and are only found in small numbers or as isolated plants. Except for geographical isolation, prezygotic isolating mechanisms (Stace, 1975) are not well developed. Co-occurring taxa occupy more or less the same ecological niche. The flowers of all taxa of *Monanthes* are very similar, and the unspecialized nectar presentation points to pollination by a generalized type of pollinators such as Diptera. Consequently isolation mechanism involving pollen vectors are not present. There is a certain seasonal isolation as flowering periods differ among sections. The species of *M.* sect. *Petrophyllae* (*M. brachycaulos, M. pallens, M. lowei*) flower first, followed by those of *M.* sect. *Sedoidea* (*M. anagensis, M. laxiflora*) and later by those of *M.* sect. *Monanthes* (*M. muralis, M. polyphylla*). But this phenological separation is incomplete, and very often some plants are found with an overlapping flowering period.

The hybridization experiments revealed that postzygotic isolation mechanisms, such as seed inviability and reduced vigour of the F1, are by and large absent in *Monanthes.* As there is no genetic isolation to prevent hybridization, hybrids can be expected to be produced wherever two or more taxa occur together.

The distribution of the perennial taxa of *Monanthes* on the Canary Islands is summarized in Table 3. Based on this compilation 18 different combinations of interbreeding taxa could be expected if all taxa from the same island would occur together somewhere, but the rather restricted distribution of some taxa on Tenerife (*M. anagensis, M. minima, M. pallens*) rule out some combinations. 12 of the 18 potentially hybrid producing combinations are discussed hereafter.

Table 3. Distribution of the perennial taxa of *Monanthes* on the Canary Islands (Nyffeler 1992). The Canary Islands are Lanzarote (LAN), Fuerteventura (FUE), Gran Canaria (CAN), Tenerife (TEN), Gomera (GOM), La Palma (PAL) and Hierro (HIE).

	LAN	FUE	CAN	TEN	GOM	PAL	HIE
M. anagensis				x			
M. brachycaulos			x	x			
M. laxiflora	x	x	x	x	x		
M. minima				x			
M. muralis						x	x
M. pallens				x	x		
M. polyphylla ssp. *amydros*					x		
M. polyphylla ssp. *polyphylla*			x	x		x	

1. *M. anagensis* & *M. laxiflora* (*M.* x *anagiflora*) are reported to grow together on the Anaga ridge of Tenerife. Several authors (Praeger, 1929; Lems & Holzapfel, 1974) recorded the occurrence of hybrids and backcrosses between these two species. Artificial F1 hybrids were obtained, but it turned out that these two taxa do not very easily produce seeds when crossed. An obvious explanation for this observation may be the significant difference in ploidy level (diploid versus hexaploid) between the two specimens used in the experiments.

2. *M. anagensis* & *M. brachycaulos* are growing together at some locations in the Anaga region on Tenerife. No intermediate plants are known, which is not surprising as the two taxa have different flowering periods.

3. *M. brachycaulos* & *M. laxiflora* (*M.* x *tilophila*) are growing together at several locations along the Northern coast of Tenerife (Punta del Hidalgo, Realejo, El Palmar) and on Gran Canaria. Hybrids and alleged backcrosses are frequently found, and the reports are confirmed by the experiments. The hybrid between these two taxa was first described as an independent species from Gran Canaria. The hybrid binominal *M.* x *chamorgensis* (nom. inval., Art. H5.1) is given for *M. brachycaulos* var. *adenopetala* x *laxiflora.* The present author does not consider *M. brachycaulos* var. *adenopetala* to be different from *M. brachycaulos* [var. *brachycaulos*].

4. *M. brachycaulos* & *M. minima* are growing together near Igueste de San Andrés on Tenerife. Even though the two taxa flower by and large at the same time, no intermediate plants have been reported. Both species belong to *M.* sect. *Petrophyllae* and are

very similar in habit. Therefore it is expected that hybrids are rather difficult to detect.

5. *M. brachycaulos* & *M. pallens* (*M.* x *hybrida*) are reported to produce hybrids at several localities on Tenerife at Anaga and the region near Los Silos (Praeger, 1929, 1932). Whereas it is unknown to the present author that *M. pallens* occurs in the Anaga, the distribution areas of the species approach one another in the region of Buenavista (Teno). If there are locations where the two taxa grow together, hybrids have to be expected, especially as they could be crossed experimentally very easily and produced a highly fertile offspring. The hybrid binominal *M.* x *pumila* is given for *M. brachycaulos* x *silensis*. The present author places *M. silensis* in the synonymy of *M. pallens*.

6. *M. brachycaulos* & *M. polyphylla* ssp. *polyphylla* (*M.* x *intermedia*) are growing together at several localities along the Northern coast of Tenerife (Punta del Hidalgo, Realejo Alto). On Gran Canaria *M. polyphylla* ssp. *polyphylla* is very local or even extinct. The phenological separation of these two species is well developed so that production of hybrids is expected to be a rather rare. Moreover, the triploid F1 hybrids look rather similar to *M. polyphylla* ssp. *polyphylla* and therefore hybrids are not easily recognized.

7. *M. laxiflora* & *M. pallens* (*M.* x *sventenii*) are growing together at several localities on Tenerife and Gomera. Hybrids and backcrosses are regularly found. These findings are supported by the crossing experiments, as the F1 hybrids proved to be rather highly fertile. Often as a third species *M. polyphylla* (on Tenerife *M. polyphylla* ssp. *polyphylla* and on Gomera *M. polyphylla* ssp. *amydros*) are found at these locations as well. In these cases the patterns of intermediate plants are so complex that it is not possible to hypothesize the involved taxa. The hybrid binominal *M.* x *burchardii* is given for *M. laxiflora* x *silensis*. The latter species is regarded conspecific with *M. pallens* by the author.

8. *M. laxiflora* & *M. polyphylla* ssp. *amydros* (*M.* x *gomerensis*; proposed for *M. amydros* (nom. inval., Art. 37.1) x *M. laxiflora* var. *laxiflora*) are growing together at several localities on Gomera (Barranco de San Sebastian). Often *M. pallens* is found there as well.

9. *M. laxiflora* & *M. polyphylla* ssp. *polyphylla* (*M.* x *isabellae*) are growing together at several locations on Tenerife (Punta del Hidalgo, Realejo Alto, Buenavista). Occasionally *M. brachycaulos* or *M. pallens* are present at these sites as well. On Gran Canaria no locations are known where these two taxa grow together.

10. *M. muralis* & *M. polyphylla* ssp. *polyphylla* are found on La Palma. Herbarium specimens studied by the author in the course of a taxonomic revision of *Monanthes* (Nyffeler, 1992) proved to be a complex mixture of both taxa and intermediate specimens, which are assumed to be of hybrid origin. The two taxa where nowhere found growing together during a visit to this island.

11. *M. pallens* & *M. polyphylla* ssp. *amydros* are found growing together at several locations on Gomera. Their flowering periods seem rather well separated, which to great extent prevents hybridization between these two taxa. At most sites *M. laxiflora* is present as well.

12. *M. pallens* & *M. polyphylla* ssp. *polyphylla* (*M.* x *silophylla*) are found growing to-

gether in the Teno of Tenerife. As mentioned above, the seasonal isolation mechanism will prevent cross-pollination in most cases.

A special case is the alleged hybrid between *M. laxiflora* and *M. muralis* from Hierro, as *M. laxiflora* is not known to occur on that island. *M. muralis* is found on La Palma and Hierro, whereas *M. laxiflora* occurs on all other five islands of the Canaries. Nevertheless a plant found in a collection of *M. muralis* from Hierro (Kuschel 90-5) strongly resembles the artificial hybrid obtained between the two species. This plant is tetraploid ('t Hart & Nyffeler, unpublished data).

The nature of *M. subcrassicaulis* is obscure (Nyffeler 1992). This name has been applied to plants somewhere intermediate between *M. muralis* and *M. polyphylla* (Praeger, 1932; Bramwell, 1971; Lems & Holzapfel, 1974). The hybrids obtained from artificial crosses between *M. laxiflora* and *M. polyphylla* ssp. *polyphylla* look very similar to the drawing of *M. subcrassicaulis* in Praeger's publication (Praeger 1932). It is assumed here that this name *M. subcrassicaulis* predominantly refers to plants of hybrid origin. *M. laxiflora* and *M. polyphylla* grow together on Tenerife and Gomera, two islands from which *M. subcrassicaulis* is reported. In addition, Praeger (1929) described the hybrid of *M. polyphylla* and *M. subcrassicaulis* (*M.* x *polycaulis*).

Although hybridization occurs at most localities with two or more taxa growing together, no breakdown of the parental taxa is ascertainable. No indications were found by the author at the localities so far studied, that hybrid swarms are formed. Nevertheless, some localities with two and more sympatric taxa of *Monanthes* on Tenerife and Gomera contain intermediate plants which cannot be simply explained as F1 hybrids. This is most obvious at locations where *M. brachycaulos* and *M. laxiflora* or *M. pallens* and *M. laxiflora* grow together. As the hybrids of these diploid species have high pollen fertilities it is to be expected that they are able to interbreed or backcross with the parents or even cross with other hybrids.

Introgression is the repeated backcrossing of a hybrid to one or both parents. The hybrids resulting from these crosses resemble the parents very closely after a few generations. This infiltration of genetic material from one species to another often occurs when the fertility of the F1 hybrids is low, obviously because it is more likely to backcross with a fully fertile parent than with another only partly fertile hybrid to produce an F2 offspring. Even though there is no obvious indication that introgression is a common phenomenon in *Monanthes*, a definite answer is not possible at the present stage.

Naming of the hybrids - The question to what extent hybrids should be named, described and keyed out in *Monanthes* accompanies the author's work with this genus since its beginning. It is important to point out that hybridization has to be taken into consideration when trying to identify a specimen. Disregard of this advice undoubtedly leads to confusion. At the present stage of knowledge it is not possible to incorporate the many different forms of hybrid plants into a key to the taxa of *Monanthes*.

The International Code of Botanical Nomenclature (Greuter & al. 1988) provides two different ways to name hybrids, either by hybrid formulas or by binominal names (= nothotaxa). Nothotaxa are marked with a multiplication sign directly before the epithet or with the prefix term "notho". They must conform with the rules of the Code, so they have to be validly published and typified. Hybrid formulae consist of the names of the parental taxa with a multiplication sign in between. Complete lists of proposed hybrid formulae and nothotaxa are given by Nyffeler (1992) and Heath (1994).

The advantages and disadvantages of both methods of naming have been comprehensively discussed by several authors (e.g. Little 1960). In *Monanthes* the present author strongly recommends the use of formulas, as the hybrids are not found outside the distribution area of their parents (i.e., the hybrids do not represent 'independent' taxa) and are of no horticultural value. Nothotaxa burden the nomenclature of the genus especially when the phenomenon of hybridization is still subject of investigation. They cause problems when the parents are given different taxonomic ranks by different authors as the ICBN demands the nothotaxa to be of the same rank as the parent with the lower rank. Furthermore, the use of hybrid formulas has the advantage of the possibility to deal adequately with introgressants. These hybrids forms can be designated with the symbols ">" and "<" instead of "x" (Hardin 1958), though the use of these symbols is not sanctioned by the ICBN.

Acknowledgements

I am greatly indebted to Dr. Henk 't Hart and Herman van der Klis of the Department of Plant Ecology and Evolutionary Biology of Utrecht University for the data on chromosome numbers and for the help with the interpretation of the results. Furthermore, my thanks go to all persons who supported my studies in *Monanthes* with live plant material.

References

Bañares Baudet, A. 1990. Híbridos de la familia Crassulaceae en las Islas Canarias. Novedades y datos corológicos II. *Vieraea* **18**: 65-85.

Bramwell, D. 1971. Studies in the Canary Islands flora: the vegetation of Punta de Teno, Tenerife. *Cuad. Bot. Canaria* **11**: 4-37.

Burchard, O. 1929. Beiträge zur Ökologie und Biologie der Kanarenpflanzen. *Biblioth. Bot.* **98**: 1-262.

Greuter, W. (ed.). 1988. International code of botanical nomenclature. Koeltz Scientific Books, Königstein.

Hardin, J.W. 1958. The annotation of introgressants. *Taxon* **7**: 52-53.

Hart, H. 't. 1987. Natural hybrids in *Sedum* (Crassulaceae). 1. Two new hybrids of *S.* series *Rupestria* and a new locality of *S.* x *brevierei*. *Bot. Jahrb. Syst.* **109**: 1-16.

Heath, P.V. 1994. A synopsis of *Monanthes* Haworth. *Calyx* **4**: 92-93.

Jacobsen, H. & Rowley, G.D. 1973. Some name changes in succulent plants. *Natl. Cact. Succ. J.* **28**: 4-6.

Lems, K. & Holzapfel, C. 1974. Flora of the Canary Islands. The Cruciferae, the Crassulaceae and the ferns and their allies. *Anales Inst. Nac. Invest. Agrar., ser. Prod. Veg.* **4**: 165-273.

Little, E.L. 1960. Designating hybrid forest trees. *Taxon* **9**: 225-231.

Mes, T.H.M. & Hart, H. 't. 1994. *Sedum surculosum* and *S. jaccardianum* (Crassulaceae) share a unique 70 bp deletion in the chloroplast DNA trnL (UAA) - trnF (GAA) intergenic spacer. *Pl. Syst. Evol.* **193**: 213-221.

Nyffeler, R. 1992. A taxonomic revision of the genus *Monanthes* (Crassulaceae). *Bradleya* **10**: 49-82.

Praeger, R.L. 1929. Semperviva of the Canary Islands area. *Proc. Roy. Irish Acad., Sect. B* **38**: 454-499.

Praeger, R.L. 1932. An account of the *Sempervivum* group. Royal Horticultural Society, London.

Radford, A.E., Dickison, W.C., Massey, J.R. & Bell, C.R. 1974. Vascular plant systematics. Harper & Row, New York.

Stace, C.A. 1975. Hybridization and the flora of the British Isles. Academic Press, London.

Stanley, R.G. & Linskens, H.F. 1974. Pollen: Biology, biochemistry, management. Springer Verlag, Berlin.

Uhl, C.H. 1961a. Some cytotaxonomic problems in the Crassulaceae. *Evolution* **15**: 375-377.

Uhl, C.H. 1961b. The chromosomes of the Sempervivoideae (Crassulaceae). *Amer. J. Bot.* **48**: 114-123.

Uhl, C.H. 1992. Polyploidy, dysploidy, and chromosome pairing in *Echeveria* (Crassulaceae) and its hybrids. *Amer. J. Bot.* **79**: 556-566.

CHAPTER 6

Quantitative phytogeography, species richness, and evolution of American Crassulaceae

Joachim Thiede

Abstract

The quantitative phytogeography of Northern American Crassulaceae is dealt with in detail, including: 1) maps indicating the overall species number and the number of species for the genera *Sedum*, *Echeveria*, and *Dudleya* per degree square; 2) species richness along an altitudinal and latitudinal gradient; 3) average area size and average latitudinal and longitudinal range in temperate and tropical regions and along a latitudinal gradient; 4) species turnover along a latitudinal gradient; 5) estimations for the spatial floristic diversity; 6) species abundance in different vegetation types; 7) delimitation of regional centres of species richness and endemism; 8) indices of similarity; 9) chorograms for these regional centres. Based on a cluster analysis, three phytogeographical regions are delimited within Northern American Crassulaceae (Eastern Temperate, Western Temperate, Tropical Mexican), which differ systematically, phylogenetically, caryologically, in species richness per degree square, overall species number and number of genera and infrageneric taxa, average area size of its species, area index and spatial floristic diversity. Based on phylogenetic data inferred from chloroplast DNA sequences, six independent introductions belonging to three infrafamilial taxa can be distinguished within American Crassulaceae. Four introductions are of minor importance with regard to the number of species, one includes *Dudleya* and *Sedum* subgen. *Gormania*, and the most important apparently comprises all or almost all tropical Mexican Crassulaceae of the *Echeveria* group. Possible routes of migration and patterns of evolutionary diversification in American Crassulaceae are outlined, especially with regard to differences between temperate and tropical regions.

Introduction

The American Crassulaceae form a major phytogeographical assemblage of three different infrafamilial taxa with thirteen genera (Table 1): 1) The subfamily Crassuloideae is represented by 12 species of the nearly cosmopolitan *Tillaea*, which mainly contains small, annual, semi-aquatic, hardly succulent plants. 2. Subtribus Telephinae (subfamily Sedoideae tribus Sedeae) is represented by 4 species in the chiefly Asian genera *Rhodiola* and *Hylotelephium*. 3. Subtribus Sedinae includes the majority of American Crassulaceae (about 96% of the species) and all genera of the former subfamily Echeverioideae. The Sedinae are represented with two large genera (*Sedum*, *Echeveria*), two medium-sized (*Villadia*, *Dudleya*) and 6 smaller genera. Except for *Sedum*, the genera of the Sedinae are endemic to America.

The distribution patterns, centres of species richness and endemism as well as the evolution of American Crassulaceae are still largely unknown. To fill this gap, the quantitative phytogeography, possible migrational pathways, and evolutionary patterns are dealt with in more detail in this paper.

Table 1. Infrafamilial classification, number of species, generic endemism, and phylogenetic relationships of the American genera of Crassulaceae.

genera intrafamilial classification ('t Hart, this volume)	number of species	endemic genera America (A)	 Mexico(M)	subfamily (Berger, 1930)	clade(s) (van Ham, 1994, this volume
Crassuloideae	12				
Tillaea	12			Crassuloideae	*Crassula*
Sedoideae-Sedeae-Telephinae	4				
Hylotelephium	1			Sedoideae	*Telephium*
Rhodiola	3			Sedoideae	*Telephium*
Sedoideae-Sedeae-Sedinae	391				
Sedum	139			Sedoideae	*Acre/Leucosedum*
Echeveria	127	A		Echeverioideae	*Acre*
Villadia	45	A		Echeverioideae	*Acre*
Dudleya	42	A		Echeverioideae	*Leucosedum*
Graptopetalum	13	A		Echeverioideae	*Acre*
Pachyphytum	13	A	M	Echeverioideae	*Acre*
Lenophyllum	7	A		Sedoideae	-
Thompsonella	3	A	M	Echeverioideae	-
Diamorpha	1	A		Sedoideae	-
Tacitus	1	A	M	-	-
all genera	407	9	3		

Material and methods

The phytogeographical analysis is completely based on extensive data collected from literature. The major sources of reference were Clausen (1959, 1975, 1977, 1978, 1979, 1981, 1984), Denton (1982), Fröderström (1935), Moran (1994), Uhl (1970, 1976b, 1978, 1980, 1983, 1985, 1992a), Uhl & Moran (1973), and Walther (1972), as well as a series of minor articles of the above authors and papers of Bartel, Britton & Rose, Bywater & Wickens, Calderón de Rzedowski, Cházaro B., Glass, Kimnach, MacDougall, Meyrán G., Moran, von Poellnitz, Praeger, Sánchez-Mejorada and others. Each of the approximately 7,000 records includes the name of the taxon (see below), the locality and all available additional data, such as infrageneric placement, habitat description, altitude or altitudinal range, collector, collection number and date, and literature reference. Since most localities are based on specimens cited in taxonomic treatments they were regarded as reliable. To exclude misidentifications, all records clearly situated outside the known distribution area were checked carefully and eventually excluded from the analyses, depending on the reporting authority.

The geographical coordinates for all localities were determined with detailed topographic maps. Records not positively associable with a given degree square were omitted from subsequent analyses. Degree squares are referred to by their respective latitudes and longitudes and major towns or regions (e.g. 18-19/97-98 = Tehuacán). Methods for the processing of phytogeographical data are exemplified in the respective chapters. The genus *Tillaea* is omitted from most analyses (except Tables 1, 2, 10) due to a lack of data.

Since no modern taxonomic account of American Crassulaceae exists, a checklist was compiled as a basis for all analyses. Major sources of reference: *Sedum* is based on Clausen (l.c.) and Uhl (l.c.), and adapted after Denton (1993); *Echeveria* is principally based on Walther (1972), but includes many corrections and additions published by Kimnach, Meyrán, Moran and Uhl; *Dudleya* is based on Bartel (1993; US species) and Moran (1981; Mexican species), including later additions.

Overall patterns of phytogeography, species richness, and evolution in American Crassulaceae

Species richness and endemism in American Crassulaceae: An overview - With about 407 species America is the continent with the greatest diversity of Crassulaceae (Table 1). Mexico has the highest number of Crassulaceae species on a global scale with 305 species (75 % of all American species; Table 2). South Africa comes second with about 225 species (Tölken, 1985). In America the USA come second in species diversity with 77 species, and Central and South America comprise only 18 and 38 species, respectively (Table 2). In Mexico the percentage of endemic Crassulaceae species is 89.8% and for Echeveria this is even 97.1% (Table 2). These figures probably represent the highest degree of endemism within large Mexican plant families and genera. The former figure by far exceeds the general estimate for the endemism on the speterminal branch in the monophyletic *Acre*-clade together with *Graptopetalum paraguayense* (N.E. Brown) E. Walther, *Pachyphytum compactum* Rose, *Villadia batesii* (Hemsl.) Baehni & Macbr., *Sedum cockerellii* Britt. (*S.* sect. *Cockerellia*), *S. bourgaei* Hemsl. (*S.* sect. *Fruticisedum*), *S. dendroideum* Moc. & Sessé (*S.* sect. *Pachysedum*), *S. reptans* R.T. Clausen (*S.* sect. *Sedum*), and, most remarkably, *S. fusiforme* Lowe from Madeira and the European *S. alpestre* Vill. (van Ham, 1994, this volume, 't Hart, this volume).

6) In contrast to previous assumptions, *Dudleya* most certainly does not belong to the *Echeveria* group within the *Acre*-clade, but forms a branch within the *Leucosedum*-clade together with *Sedum moranii* R.T. Clausen of *S.* subgen. *Gormania* ('t Hart, this volume, unpublished results). It thus appears that *Dudleya* evolved independently from the genera of the *Echeveria* group, and that Berger's (1930) subfamily Echeverioideae is polyphyletic and unnatural (van Ham, 1994, this volume). The origin of *Dudleya* independent from the *Echeveria* group agrees with its apparent reproductive isolation (Uhl, 1994a) as well as a number of morphological differences (Moran, 1942; Thomson, 1993). Furthermore, the seed surface ornamentation of *Dudleya* and *Sedum* subgen. *Gormania* differs from the *Echeveria* group (Knapp, 1994).

Delimitation of phytogeographical regions and overall patterns of distribution, species richness, and evolution - In order to delimit exactly quantified phytogeographical units, cluster analysis was performed with the programme "SPSS for Windows" version 6.0 (Brosius, 1989). The analysis contained, in the terminology of Crovello (1981), the single degree squares as "Operational Geographic Units" (OGUs) which together form the "Operational Geographic Set" (OGS), and the different taxa as "characters" whose presence or absence within the OGUs represent the "character states". Two different analyses were performed. In the first analysis the species were used as characters (in order to limit the data set species reported from only one grid were excluded). The second analysis included all genera and infrageneric groups listed in Table 3 (excluding *Sedum* sect. *Sedum*, which is an unnatural hold-all of species not assignable to any more well defined section). The data matrix included 853 OGUs x 341 characters at the species level and 853 OGUs x 50 characters at the generic/infrageneric level. The delimitation of phytogeographical units is predominantly based on the cluster analysis at the generic/infrageneric level and corrected intuitively for artifacts due to degree squares with low species numbers. The resulting phytogeographical units and subunits fit reasonably well into the hierarchic phytogeographical system of Takhtajan (1986) used here for comparison (Takhtajan's phytogeographical units are placed in quotation marks).

The distribution of Northern American Crassulaceae is by and large restricted to the major mountain ranges, viz. the Appalachians and adjacent plateaus in the Eastern USA, the Rocky Mountains and various coastal mountain ranges in the Western USA and Western Canada, and the various mountain ranges and highlands in Mexico and Guatemala. Three major phytogeographical regions can be distinguished in this area, which widely correspond to the Western and Eastern temperate parts of the USA and Canada and to the (sub)tropical parts of Mexico. They are further divided into subunits which by and large agree with phytogeographical (sub-)provinces of Takhtajan's system (1986). The delimitation of these three regions and the number of species per degree square are shown in Figure 1 and 2. The number of species and number of endemic species of each region are presented in Table 3. The high degree of distinctiveness at the generic and infrageneric level and the high degree of endemism on the species level within each region (72.7 - 97.5%) and its subunits (72.7 - 82.1%) clearly demonstrate that these phytochoria represent well defined natural entities (Table 3, 4). The phytogeographic (Fig. 1, 2), systematic (Table 1), and phylogenetic distinctness of these three phytochoria indicates long periods of separation and by and large independent evolutionary histories.

1) The Eastern Temperate Crassulaceae Region is largely congruent with the "North American Atlantic Region" (Fig. 1) and encompasses Eastern temperate North America (1A) and the Atlantic parts of the "Circumboreal Region" (1B). Overall number of species (11), number of infrageneric taxa (7), maximum species richness per degree square (4), and altitudinal range are clearly the lowest of all three regions (Fig. 1, 6A, Table 4).

The Eastern Temperate Region includes introductions of *Hylotelephium*, *Sedum* sect. *Subrosea*, *Rhodiola* subgen. *Rhodiola*, and *Tillaea*, of which the first two are endemic. The phylogenetic position of *Diamorpha*, *Sedum* sect. *Ternata*, and *Sedum* sect. *Tetrorum* is unknown. The close relationship of some of the American and Eurasian species (*Hylotelephium*) or even specific identity (*Rhodiola*, *Sedum* sect. *Subrosea*) suggests that these groups were quite recently introduced. Clausen (1975) assumed a common ancestor for the closely related *Hylotelephium telephioides* (Michx.) H. Ohba from Eastern North America and the Eurasian *H. telephium* (L.) H. Ohba., and supposed that the ancestral area was split during the glacial periods.

Eastern temperate Crassulaceae are centred in and around the Appalachian Mountains ("Appalachian Province"), but sparsely represented in the Atlantic and absent from the Gulf coastal plains ("Atlantic and Gulf Coastal Plain Province") and the Great Plains ("North American Prairies Province"). Two species, *Sedum villosum* L. and *Rhodiola rosea* (L.) Scop., are (mainly) restricted to the Eastern coastal Atlantic parts of the "Circumboreal Region" North of 41°(1B in Fig. 1).

2) The Western Temperate Crassulaceae Region encompasses Western temperate North America and is geographically and systematically completely distinct from the Eastern Temperate Region (Fig. 1, Table 3) . The clear separation from the preceding Eastern Temperate Region may be due to the intermittent Great Plains, whose expressed continental climate and lack of mountains and rocky habitats could have prevented migration and floristic exchange.

The Crassulaceae of the Western Temperate Region belong to at least 4 independent introductions, i.e. *Dudleya* (endemic), *Sedum* subgen. *Gormania* (endemic), *Rhodiola* subgen. *Clementsia* (near-endemic), *R.* subgen. *Rhodiola*, and *Tillaea*. The phylogenetic position of *Sedum* sect. *Lanceolata* and *S.* sect. *Sedella* is not known. Since the majority of the Western temperate and some Eastern temperate species have

Table 3. The number of species and number of endemic species for all genera and infrageneric taxa of American Crassulaceae within the three Northern American Crassulaceae regions and their subu-

		East.Temp.R.				Western Temperate Region							
genus / infrageneric taxon / informal species group	species number	Circumboreal Region	eastern part	North American	Atlantic Region	Circumboreal Region	western part	Rocky Mountain	Province	Vancouverian Prov. /	Californian Province	Baja Californian	Subprovince
	spp.	all	end.	all.	end.	all	end.	all	end.	all	end.	all	end.
Sedum subg. Sedum sect. Subrosea `t Hart	1	1	0										
Hylotelephium H.Ohba	1			**1**	**1**								
Diamorpha Nuttall	1			**1**	**1**								
Sedum subg. Sedum sect. Ternata Berger	4			**4**	**4**								
Sedum subg. Sedum sect. Tetrorum (Rose) R.T.Clausen	1			**1**	**1**								
Rhodiola Linnaeus subg. Rhodiola sect. Rhodiola	2	2	0	1	0	1	0	1	0	1	0		
Sedum subg. Sedum sect. Lanceolata R.T.Clausen	9			1	1			5	2	6	2		
Sedum subg. Gormania sect. Rosulata (Bgr.) Clausen & Uhl	1							1	0	1	0	1	0
Sedum subg. Gormania (Britton) Clausen sect. Gormania	8									**8**	**8**		
Sedum subg. Gormania sect. Oreganica R.T.Clausen	1									**1**	**1**		
Sedum subg. Sedum sect. Sedella (Britton & Rose) Berger	4									**4**	**4**		
Dudleya Britton & Rose subg. Dudleya	29									6	2	27	21
Dudleya subg. Hasseanthus (Rose) Moran	4											**4**	**4**
Dudleya subg. Stylophyllum (Britton & Rose)	9											**9**	**9**
Rhodiola subg. Clementsia (Rose) H.Ohba	1							1	0				
Sedum subg. Sedum sect. Cockerellia R.T.Clausen & Uhl	2							1	0			1	1
Sedum subg. Sedum group Alamosanum Fröderström	13											1	0
Echeveria ser. Thyrsiflorae Moran	3												
Echeveria ser. Urceolatae E. Walther	14												
Sedum subg. Sedum sect. Pachysedum Berger	19												
Sedum subg. Sedum sect. Sedastrum (Rose) Berger	4												
Sedum Linnaeus subg. Sedum sect. Sedum	24												
Villadia sect. Altamiranoa (Rose) R.T.Clausen	24												
Villadia Rose sect. Villadia	20												
Echeveria ser. Pruinosae E. Walther	4												
Echeveria ser. Secundae (Baker) Berger	7												
Echeveria ser. Spicatae (Baker) Berger	4												
Graptopetalum sect. Byrnesia (Rose) Moran	6												
Sedum subg. Sedum group Moranense Fröderström	13												
Sedum subg. Sedum sect. Sedum Sedum palmeri-group Uhl	3												
Echeveria ser. Mucronatae E. Walther	3												
Graptopetalum Rose sect. Graptopetalum	7												
Sedum subg. Sedum sect. Fruticisedum Berger	14												
Echeveria ser. Angulatae E. Walther	8												
Echeveria ser. Ciliatae Moran	4												
Echeveria De Candolle ser. Echeveria	6												
Echeveria ser. Gibbiflorae (Baker) Berger	17												
Echeveria ser. Nudae E. Walther	24												
Echeveria ser. Racemosae (Baker) Berger	21												
Thompsonella Britton & Rose	3												
Pachyphytum Link, Klotzsch & Otto sect. Pachyphytum	7												
Sedum subg. Sedum group Napiferum Fröderström	8												
Echeveria ser. Occidentales Moran	2												
Tacitus Moran & Meyrán	1												
Pachyphytum sect. Diotostemon (Salm-Dyck) E. Walther	3												
Pachyphytum sect. Ixiocaulon Moran	2												
Echeveria ser. Valvatae Moran	2												
Echeveria ser. Induplicatae Moran	1												
Sedum subg. Sedum sect. Centripetalia Alexander	2												
Echeveria ser. Paniculatae Berger	1												
Echeveria ser. Chloranthae Moran	1												
Lenophyllum Rose	7												
species	380	3	0	9	8	1	0	9	2	27	17	43	35
% endemic species				88,9				22,2		63,0		81,4	

nits as well as in Central and South America, which principally belong to the Tropical Mexican Region (doubtful records were excluded from the analysis).

Tropical Mexican Region								Centr. Amer.				South Amer.			
Sierra Madre Occi-	dental Subprovince	Trans-Mex. Volc. Belt	& Mex. Altipl. Subpr.	Sierra Madre d. Sur &	mts.Carib. Reg. Prov.	Sierra Madre Orien-	tal Subprovince	Guatemala		El Sal., Hond., Nicar.,	Costa Rica, Panama	Venezuela, Colombia	Ecuador	Peru, Bolivia	Chile
	end.	all	end.	all	end.	all	end.	all	end.	all	end.	all	end.	all	end.
1	0														
2	0														
1	0														
1	0					1	0								
8	4	8	5			3	0								
1	0	2	0	1	1	1	0								
1	1	9	8	1	1	4	3								
2	2	7	4	13	10	1	0	1							
1	0	4	1	2	0	2	0	1		1					
2	1	9	6	9	7	5	4	2		1				3	3
1	1	8	7	6	4	1	0							10	10
1	1	6	6	9	9	4	4	2	1						
		1	1	2	2	1	1								
		4	2	5	3	1	0								
		2	1	3	2	1	0	1		1					
		4	4	1	1	1	1								
		5	2	5	4	7	4								
		1	0	2	2	1	0								
1	0	1	0	3	2			1	1						
5	5	1	1	1	1										
		9	6	7	5	1	0	1	1						
		8	4			4	0								
		2	0	4				1							
		1	1	5	5			1	1						
		11	8	9	6			1	1						
		3	3	14	14			3	1	2	1	6	6		
		1	0	13	12			1	1			2	2	6	6
		2	1	2	1										
		6	6			1	1								
3	0	8	5												
1	1	1	1												
1	1														
		3	3												
		2	2												
		2	2												
		1	1												
				2	2										
				1	1										
				1	1										
						7	7								
33	17	132	91	123	96	47	25	16	7	5	1	8	8	19	19
51,5		68,9		78,0		53,2		43,7		20,0		10,0		10,0	

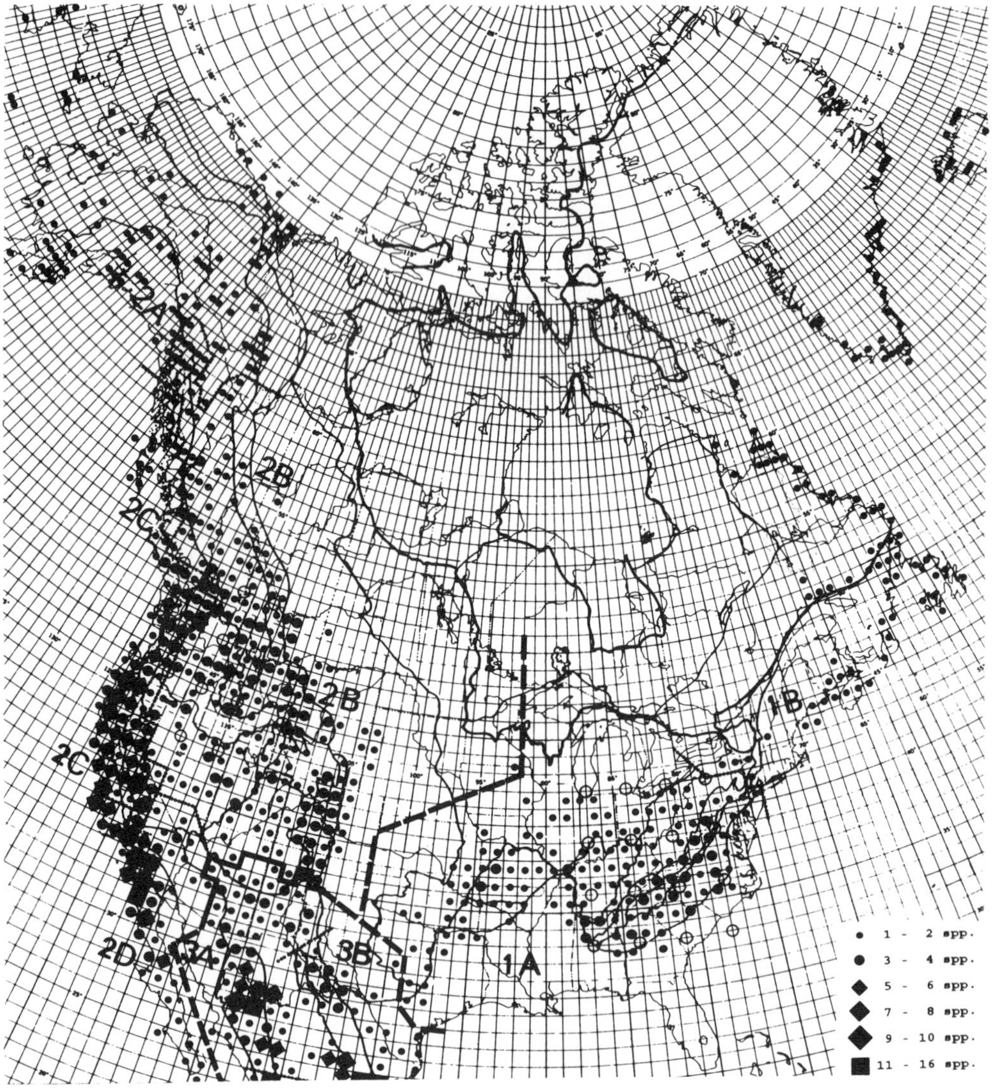

Figure 1. Distribution and number of species per degree square of Crassulaceae in Canada, the USA and Northern Mexico. The borders of the three phytogeographical Crassulaceae regions are indicated with continuous or broken lines, the borders of their subunits with dotted lines (further explanation in the text). The borders of the Sonoran and Chihuahuan Deserts (MacMahon & Wagner, 1985), the Tehuacán Valley, and the Balsas Basin are indicated by finely dotted lines. Major mountain ranges are indicated by lines, less eminent mountain ranges by broken lines. The squares used for the estimation of the spatial floristic diversity are indicated by open circles.

the same type of seed surface ornamentation (Knapp, 1994) the temperate American Crassulaceae may well represent one phylogenetic lineage, which would include 87% (69 out of 79 species) of all temperate species.

The Western Temperate Crassulaceae Region encompasses the Rocky Mountains ("Rocky Mountain Province") and various coastal ranges ("Vancouverian Province")

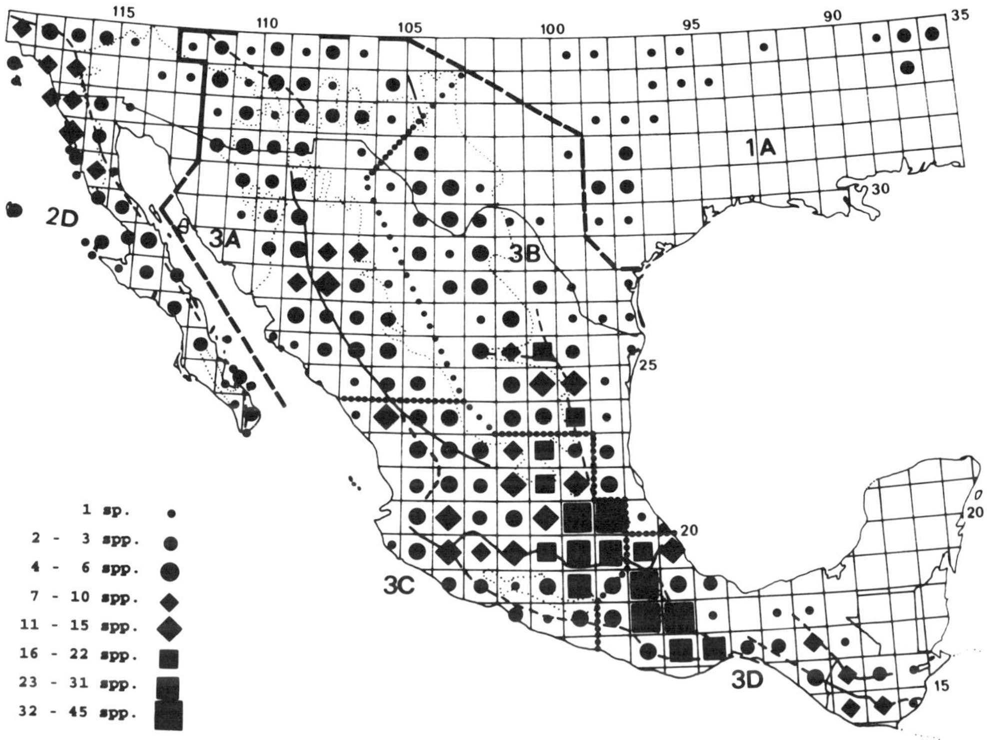

Figure 2. Distribution and number of species per degree square of Crassulaceae in the Southern USA and Northern Mexico (see Figure 1 for further explanation).

within the "Rocky Mountain Region", but also the Californian floristic region ("Californian Province"), the rest of Baja California ("Baja Californian Subprovince") and the Great Basin ("Great Basin Province") within the holarctic "Madrean Region". Overall species number (67), number of infrageneric taxa (120), maximum species richness per degree square (13), and altitudinal range are clearly higher than in the Eastern Temperate Region (Fig. 2, 6A, Table 3, 4). Four subunits can be distinguished (Fig. 1, 2):

2A) The Western part of the "Circumboreal Region" in Western Canada and Alaska only contains the widespread hemicryptophyte *Rhodiola integrifolia* and represents a counterpart to the Eastern part of the "Circumboreal Region" (1B) within the Eastern Temperate Region.

2B) The "Rocky Mountain Province" comprises the near-endemic *Rhodiola* subgen. *Clementsia*, *R. integrifolia*, and several partly endemic species of *Sedum* sect. *Lanceolata*. The latter two taxa also occur in the province described below. Overall species number (9) and the maximum number of species per degree square (5) are low (Fig. 1, Table 3). Some species extend into the "Great Basin Province", where Crassulaceae are sparsely represented. The sharp Eastern boundary coincides with the Rocky Mountains and Great Plains border.

Table 4. Number of (infra)generic taxa, species number and degree of endemism within the three Crassulaceae regions distinguished within northern American Crassulaceae.

Region	genera / infrageneric taxa			species			species within subunits endemic	species within region %
	number	endemic	%	number	endemic	%		
Eastern Temperate	7	5	71.4	11	8	72.7	8	72.7
Western Temperate	12	6	50.0	69	63	91.3	54	78.3
Tropical Mexican	40	35	87.5	279	272	97.5	229	82.1
all regions	52	46	88.4	352	343	97.4	291	82.7

2C) The "Vancouverian-Californian Province" covers the coastal ranges of most of the Vancouverian and the Northern parts of the Californian Province and is characterized by 4 endemic, annual species of *Sedum* sect. *Sedella* and by *S.* subgen. *Gormania* (Table 3). These groups, together with several species of *Dudleya,* are responsible for the high species richness (5-10 species per degree square) in Oregon and Northern and Central California (Fig. 1). This Northern part of the "Californian Province" could be delimited as a further subunit.

2D) The "Baja Californian Subprovince" includes the Baja California peninsula, the Southern part of the "Californian Province", and parts of Western Arizona in the "Sonoran Province". This subunit is characterized by two endemic subgenera of *Dudleya, D.* subgen. *Hasseanthus* and *D.* subgen. *Stylophyllum,* and many endemic species of *D.* subgen. *Dudleya* (Table 3). Many species of *Dudleya* are local endemics of the islands bordering Baja California (Fig. 3). The "Baja Californian Subprovince" has the highest number of species of all subunits (43) and the highest figure for the species richness per degree square (13) within the (Western) Temperate Region (Fig. 2, Table 3).

Dudleya shows a strong preference for the winter rainfall climate of the "Californian Province" (Fig. 3) and species richness decreases towards the North (increasing humidity and frost influence) and the South (increasing summer rainfall), respectively. Only two taxa extend into Western Arizona and one of them also into North-Western Sonora which has a biseasonal winter-summer rainfall regime. Both taxa may represent relicts of formerly more expressed winter rainfall conditions, which prevailed in the Late Wisconsinan and Early Holocene. Subsequently these taxa have been reduced in numbers or withdrew from much of the South-Western USA since the Middle and Late Holocene (van Devender & Spaulding, 1979).

Uhl (1994a), on the other hand, hypothesized *Dudleya* to have originated on the Mexican mainland sharing a common ancestor with the *Echeveria* group and to have already been present on the California land mass of West of the San Andreas Fault, which has drifted Northwards and was united with the Pacific Plate, after it had been separated from the Mexican West coast about 20 million years ago.

However, these phylogenetic speculations imply contrasting scenarios. If *Dudleya*

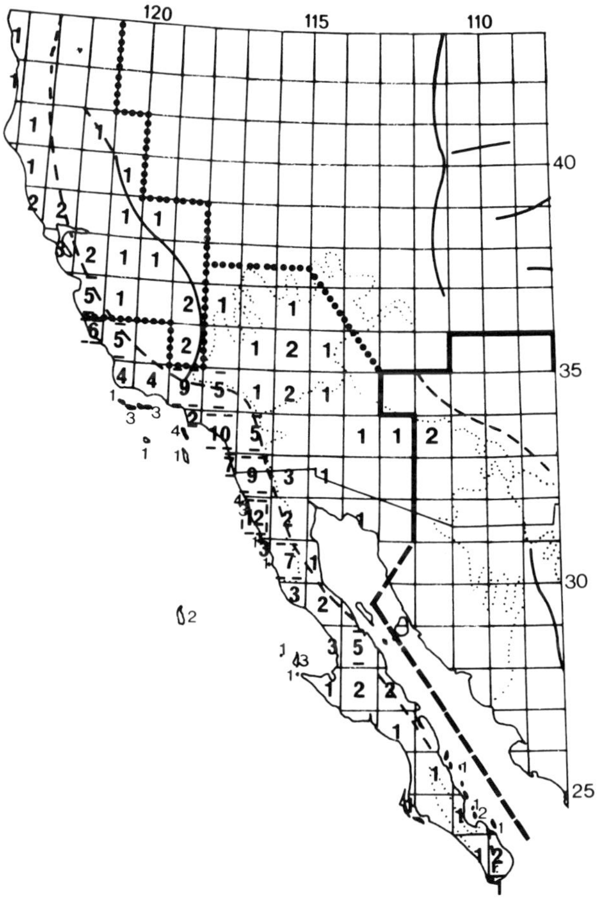

Figure 3. Total distribution and number of species per degree square of *Dudleya* in the southwestern USA and northwestern Mexico with the number of species on larger islands indicated separately (figures indicating the number of species per square are marked with bars and/or circles to emphasize differences).

evolved from *Sedum* subgen. *Gormania* in Western North America (Walther, 1936, 1972; Denton, 1982) it colonized California from the temperate North instead of the tropical South. Consequently, *Dudleya* may not have evolved before "the development of a summer-dry climate [in California] during the past million years" (Raven & Axelrod, 1978) and, in contrast to Uhl's (1994a) theory, may be of relatively recent origin.

3) The Tropical Mexican Crassulaceae Region encompasses the subtropical and tropical parts of Mexico and Guatemala and is by far the species-richest region (Table 4). With the exception of Baja California, the border between the tropical and the two temperate regions agrees approximately with the circumscription of the Holarctic and Neotropical Floristic Kingdoms.

Except for *Tillaea*, Tropical Mexican Crassulaceae belong exclusively to the *Acre*-clade (van Ham, 1994, this volume; 't Hart, this volume, unpublished data). Especially the cytological studies and hybridization experiments of Uhl (summarized in Uhl,

1992b) reveal that the genera of the *Echeveria* group, *Sedum* sect. *Pachysedum*, and to a lesser extent also other groups of Mexican *Sedum* form a single, large comparium, which most probably is the largest of the plant kingdom. The apparent absence of distinct morphological discontinuities and genetic barriers (Walther, 1972; Moran, 1978; Uhl, 1976a, 1994b) between American Sedoideae and Echeverioideae is reflected in their classification in subtribus Sedinae (Table 1; 't Hart, this volume). These findings also imply that all tropical Mexican, Central, and South American Crassulaceae (except for *Tillaea*) probably represent one single lineage, which was derived from an ancestral, Eurasian *Sedum* and comprises about 75% of all American Crassulaceae species. The common type of seed surface ornamentation of almost all species of the *Echeveria* group (Knapp, 1994) also supports this theory.

In the Tropical Mexican Region, Crassulaceae are mainly restricted to mountains and highlands (Fig. 2) which together form the "Mexican Highlands Province" within the holarctic "Madrean Region". The Tropical Mexican Crassulaceae Region includes five subprovinces, i.e. the Sierra Madre Oriental and Occidental, the Trans-Mexican Volcanic Belt, the Sierra Madre del Sur and the Southern part of the Mexican Altiplano. The concept of Rzedowski's (1983) "Región Mesoamericana de Montaña" is largely similar, but he regards this region as intermediate between the Holarctic and Neotropical Floristic Kingdoms.

The distinctiveness of the Tropical Mexican Region from the temperate regions is remarkable, since the Rocky Mountains and Sierra Madre Occidental as well as the Appalachians and Sierra Madre Oriental are more or less linked by mountain chains or plateaus which are generally regarded as important Southward migrational pathways for temperate plant groups (Rzedowski, 1983). The two most prominent examples of migration are *Pinus* and *Quercus* (Styles, 1993). In the Crassulaceae, however, there is at present no evidence for such migration.

The highest number of species per degree square in the Tropical Region (45) is found in the Tehuacán Valley (18-19/97-98), followed by the arid regions of Hidalgo-Querétaro (20-21/98-100) and the Oaxaca-Puebla border region (17-18/96-98) with 36 to 38 species. A region with at least 20 species per degree square ranges continuously from East of San Luis Potosí (21-22/100-101) up to Southeastern Oaxaca (16-17/95-96). Isolated squares with at least 10 species are situated along the Sierra Madre Oriental, around Guadalajara (19-21/103-104), at the Mazatlan-Durango highway (23-24/105-106), in the Barranca del Cobre region (27-28/107-109), in Chiapas (16-17/92-93), and in Guatemala (Fig. 2). Crassulaceae are sparsely represented in the Chihuahuan Desert, the Sonoran Desert and the warm and arid Pacific lowland regions and almost completely absent from the warm and humid lowlands of the Gulf Plains.

The overall species number (279), the number of infrageneric taxa (40), the degree of endemism at both levels (97.5 and 87.5%, respectively), and the maximal number of species per degree square (45) in the Tropical Mexican Region are clearly higher than in both temperate regions (Fig. 2, Table 4). The high number of species per degree square is only surpassed by *Crassula* in South Africa with a maximum of 61 species per degree square (Jürgens, 1991, this volume).

Sedum and *Echeveria* together constitute a considerable part of the diversity of Mexican Crassulaceae. The distribution of the two genera reflects the overall pattern of distribution and species richness of the Crassulaceae, but *Sedum* (Fig. 4) is more dominant in the Tehuacán Valley as well as in Northern Mexico and the isolated regions mentioned above, whereas *Echeveria* (Fig. 5) is clearly more abundant in the region between Hidalgo and Oaxaca.

Within the Tropical Mexican Crassulaceae Region, 4 phytogeographical subunits can be distinguished (Fig. 1, 2):

3A) The "Sierra Madre Occidental Subprovince" includes the Northern parts of that mountain range and major parts of Arizona and New Mexico in the Southern USA.

3B) The "Sierra Madre Oriental Subprovince" covers the Northern parts of that mountain range and extends Northwards into South-Western Texas.

3C) The "Trans-Mexican Volcanic Belt and Mexican Altiplano Subprovince" unites major parts of the previous two subprovinces, but also contains the Southern Sierra Madre Occidental and the Western part of the Sierra Madre del Sur.

3D) The "Sierra Madre del Sur and mountains of the Caribbean Region Subprovince" includes the mountains and highlands of that mountain range in the states of Puebla and Oaxaca and the humid Eastern parts of the Trans-Mexican Volcanic Belt, but also the mountains of the neotropical "Caribbean Region" in Chiapas and Guatemala.

The two latter subunits are richest in species and infrageneric taxa of *Sedum* and *Echeveria*, and exhibit a high degree of endemism at both levels (Table 3). Most of the 29 infrageneric taxa occur in at least two subunits of the Tropical Mexican Region and only 9, predominantly mono- or ditypic taxa, are endemic to one of the four subunits (Table 3).

The wide distribution of many infrageneric taxa of *Sedum* and *Echeveria* (Table 3) and the continuous floristic relations along the Mexican mountain ranges (Table 10) imply intensive floristic exchange between these mountain ranges including dispersal in different directions from a hypothetical major evolutionary centre in Central Mexico. Such migrations and exchanges may have taken place especially across the Neovolcanic Belt and the adjacent Southern part of the Mexican Altiplano which are intermittent between the mountains of Northern, Southern, Western, and Eastern Mexico. This assumption contrasts with Cházaro's (1992) theory, which regards the Neovolcanic Belt as an important barrier for dispersal of Crassulaceae. However, Chazaro's analysis exclusively comprised species from the far West and East of the Belt and neither the total distribution of these species nor the species occurring in between the two regions, which clearly connect both regions (Table 10), were considered. Furthermore, he restricted his analysis to the species level, whereas both regions are clearly linked at infrageneric level (Table 3). The importance of the Neovolcanic Belt for floristic exchange between Mexican mountain regions was also demonstrated in phytogeographical studies of mosses (Delgadillo, 1971, 1987).

There is almost no evidence for floristic exchange between the Sierra Madre Occidental and Oriental across the Chihuahuan Desert, despite the fact that vegetation belts during the Wisconsin-glaciation were lowered for at least 1000 m in a conservative estimation (McDonald, 1993). Of the species which have their main distribution

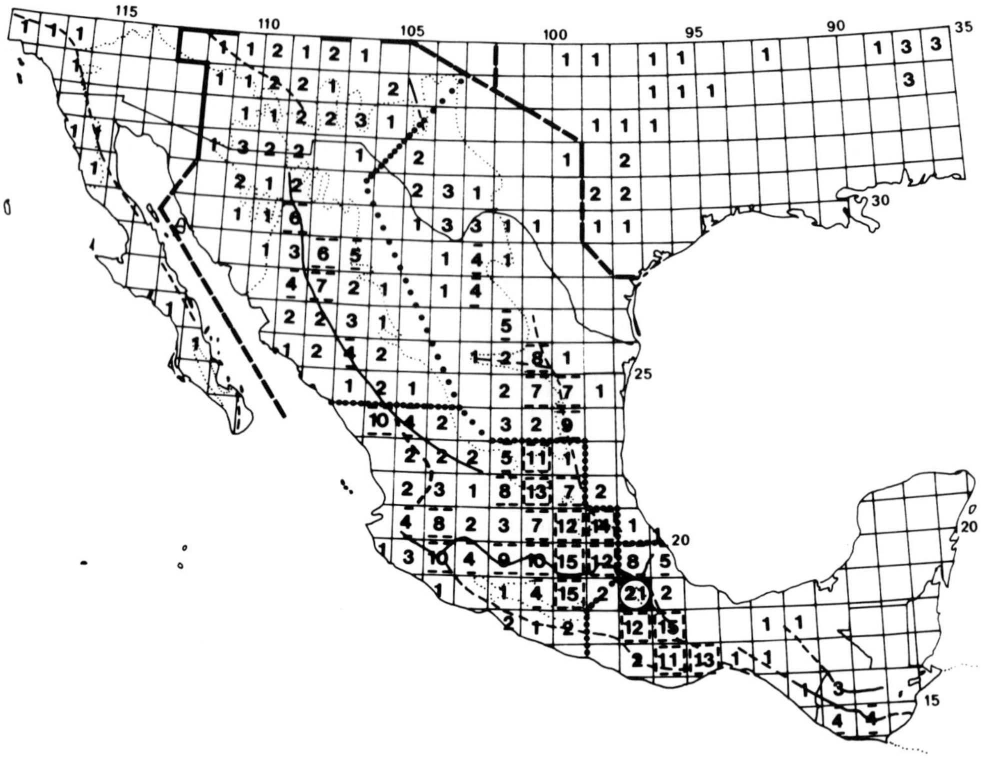

Figure 4. Distribution and number of species per degree square of *Sedum* in Mexico, Guatemala and the southern USA. (see Figure 3 for further explanation).

in the Sierra Madre Oriental only very few also occur in the Eastern Sierra Madre Occidental (Fig. 9B) as a result of migration or long-distance dispersal across the Chihuahuan Desert. Crassulaceae species are never widely distributed in both sierras.

In Arizona *Sedum* species are restricted to the Eastern summer rainfall parts (Fig. 4) in contrast to *Dudleya* which occurs in the Western winter rainfall region (Fig. 3). These differences in the phytogeographical affinities between Eastern and Western regions represents a general pattern in the flora of Arizona (McLaughlin, 1989).

Species richness along an altitudinal gradient - The overall species richness of Northern American Crassulaceae along an altitudinal gradient is nearly constant and relatively low at lower altitudes, but increases considerably above 800 m and reaches a pronounced peak at about 1,800-2,400 m (Fig. 6A,B). At higher altitudes species richness drops steeply and only few species occur above 3,400 m. The highest locations for Crassulaceae in America are reported from Mt. Ixtaccíh Chuatl at 4,400 m (*Villadia alpina* (Fröd.) Jacobs.) and 4,325 m (*Echeveria secunda* Booth).

The altitudinal zonation of temperate and tropical Crassulaceae shows pronounced differences. The overall species richness of both temperate regions (Fig. 6A) is highest in lowland regions and decreases at a constant rate towards higher altitudes, whereas in

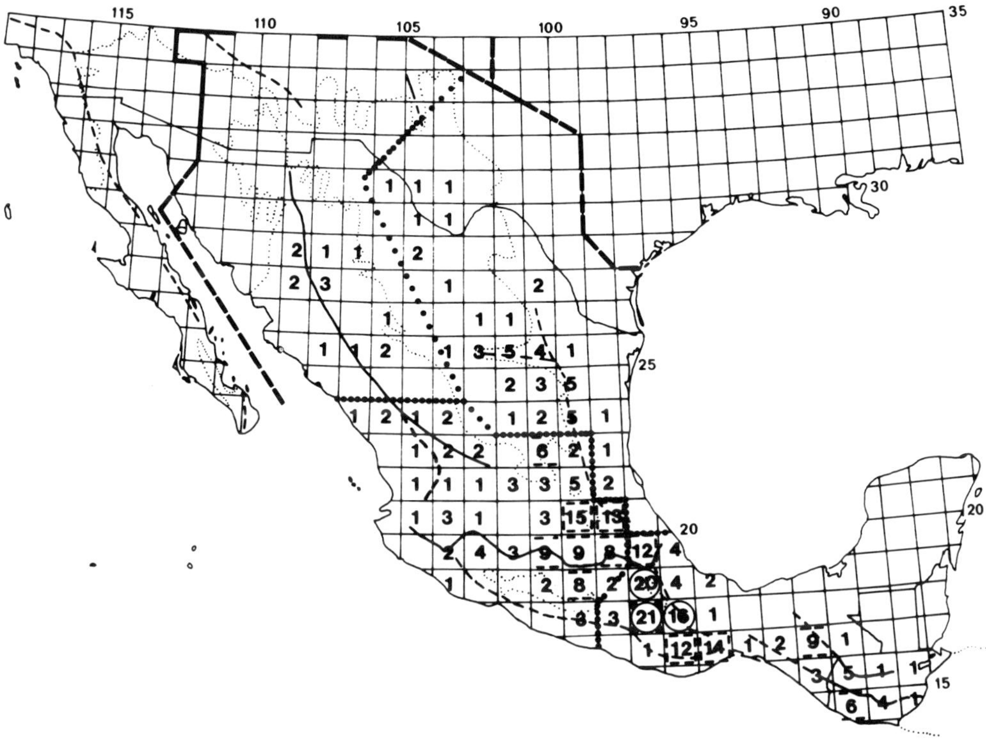

Figure 5. Distribution and number of species per degree square of *Echeveria* in Mexico, Guatemala and the southern USA. (see Figure 3 for further explanation).

the Tropical Mexican Region (Fig. 6B) higher species numbers are restricted to higher altitudes at 1,200 - 2,600 m and very few species occur in the lowlands. These differences are clearly reflected in the altitudinal zonation of *Sedum*. In both temperate regions *Sedum* is most diverse at lower altitudes, though somewhat less in the Western Temperate Region (Fig. 6A), but in the Tropical Mexican Region (Fig. 6B) it is most diverse at higher altitudes. *Echeveria* shows a pattern similar to Mexican *Sedum*, but reaches its highest diversity at somewhat lower altitudes, at 1,800 - 2,200 m (*Sedum* at 2,200 - 2,600 m), and is less diverse at altitudes below 1,500 m (Fig. 6B). The temperate *Dudleya* shows a gradually decreasing diversity towards its highest location at 3,025 m (Fig. 6A) and is the only large American genus with a preference for lowland habitats.

This difference of 1,000 - 1,500 m between the temperate and the tropical region can be interpreted with the law of "Biotopwechsel und relativer Standortkonstanz" (Walter & Breckle, 1983). Crassulaceae are principally adapted to a temperate winter or winter-summer rainfall climate (Teeri & al., 1978; Böttcher & Jäger, 1984) and their altitudinal range in tropical summer rainfall regions has shifted towards higher altitudes with the more moderate climate and higher rainfall, buffering for the higher temperatures and higher evapotranspiration during the rain period and the more pro-

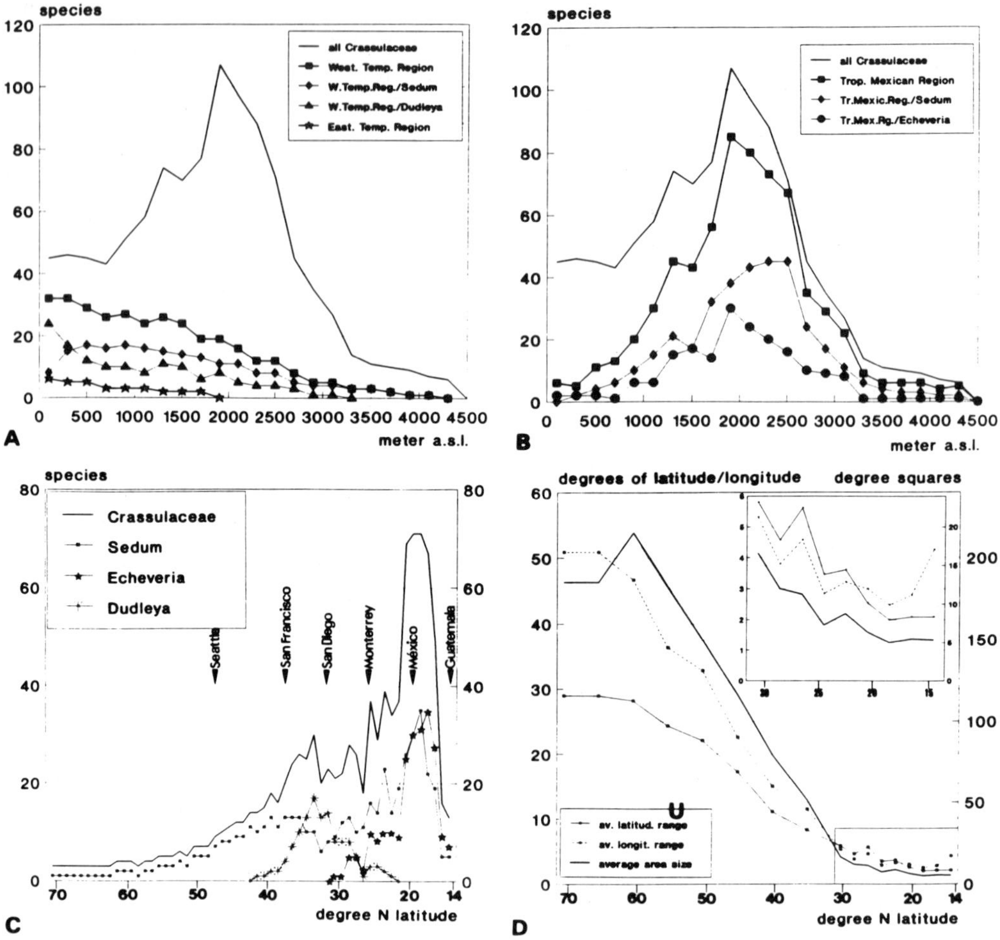

Figure 6. Aspects of species richness and area size of Northern American Crassulaceae. 6A. Altitudinal gradient of species richness at intervals of 200 m in the Western and Eastern Temperate Region. 6B. Idem in the Tropical Mexican Region. 6C. Latitudinal gradient of species richness in northern American Crassulaceae along a latitudinal gradient (latitudes of some major towns is indicated for comparision). 6D. Average latitudinal and longitudinal range and average area size along a latitudinal gradient of all species within an intervals of 1 degree of latitude (lowermost part of the curves enlarged at upper right, further explanations in the text).

nounced and longer dry period. According to a general hypsometrical gradient of 0.6 °C per 100 m (Lauer, 1993), the above altitudinal shift would compensate for a temperature raise of 6.5 - 10 °C.

Comparison of average area size and average latitudinal and longitudinal range between tropical and temperate Crassulaceae species - The latitudinal and longitudinal range (in degrees of latitude and longitude) for all Northern American Crassulaceae species was determined from the difference between the Northern- and Southern-most and the Western- and Eastern-most locality, respectively, and is regarded as continuous. The area size (ellipsoid) was calculated from both values for each species (in degree squares). In species with a non-ellipsoid area outline, the value approximates the number of degree squares encircled by the outline of the distribution area.

Average latitudinal and longitudinal range are 2.6 to 3.2-fold higher in temperate than in tropical Crassulaceae and *Sedum* species (Table 5) and area size is even 4.2-fold (all species) or 6.1-fold (*Sedum*) higher. *Sedum* shows in almost all cases values above average. Remarkably, temperate *Dudleya* has values far below average for the temperate region and nearly matches tropical *Echeveria*, which reaches the lowest values for all three parameters. In most cases average longitudinal range slightly exceeds or nearly equals average latitudinal range.

The differences in area size between temperate and tropical regions are largely due to the presence of many local endemics in the Tropical Mexican Region (Table 6). The temperate regions have a higher amount of more widespread species, and local endemics are mainly restricted to *Dudleya*. Only 39.2% (31 out of 79 species) of all temperate Crassulaceae species have an area size smaller than one degree square (predominantly *Dudleya* species), but this figure is considerable higher among tropical Crassulaceae 57.4% (155 out of 270 species). The corresponding figures in *Sedum* are in both regions below average, whereas *Dudleya* almost equals the high values for the tropical region and *Echeveria*, respectively.

Table 5. Average area size (degree squares), average latitudinal range (degrees of latitude), and longitudinal range (degrees of longitude) of the distribution areas of northern American Crassulaceae in the temperate and tropical region.

	temperate region			tropical region		
	all	*Sedum*	*Dudleya*	all	*Sedum*	*Echeveria*
number of species	79	32	42	270	97	106
area size	22.29	31.37	3.14	5.21	5.10	3.05
latitudinal range	4.49	6.08	1.77	1.70	2.12	1.31
longitudinal range	5.52	7.27	1.51	2.06	2.14	1.34

Species richness along a latitudinal gradient - Latitudinal species richness of Northern American Crassulaceae shows a pronounced increase towards tropical latitudes and reaches a maximum in Southern Mexico (Fig. 6C). The graph by and large follows an exponential curve and agrees with similar latitudinal gradients of species richness of many plants and animals (Pianka, 1966; Cox & Moore, 1993).

Latitudinal species richness of *Sedum* reflects the pattern described above and rises to a pronounced peak in Southern Mexico. The pattern of *Echeveria* in Southern Mexico is similar to *Sedum*, but species richness of *Echeveria* decreases faster in Central Mexico and even more so in Northern Mexico. *Dudleya* is most diverse in Southern California and Northern Baja California, but species richness decreases fast both Southwards and Northwards.

Table 6. Number of species of Northern American Crassulaceae within different intervals of area size (degree degrees) and comparison between temperate and tropical regions.

	N Amer.	temperate region			tropical region		
area size	all	all	*Sedum*	*Dudleya*	all	*Sedum*	*Echeveria*
> 0-0.2	140	20	6	14	118	35	49
> 0-1	188	31	7	24	155	48	63
% all species	54.5	39.2	21.9	57.1	57.4	49.5	59.4
> 1-2	27	3	-	3	24	5	13
> 2-3	15	5	1	4	10	5	2
> 3-4	14	2	1	1	12	6	5
> 4-5	15	6	3	3	9	2	5
> 3-4	12	2	1	1	18	4	5
> 6-7	8	1	1	-	7	5	1
> 7-8	6	1	1	-	5	3	1
> 8-9	5	2	1	1	4	-	1
> 9-10	8	4	1	3	5	3	2
> 10-15	11	1	1	-	10	5	4
> 15-20	9	4	3	-	5	4	1
> 20-25	7	1	-	1	7	4	1
> 25-30	5	3	1	1	2	1	1
> 30-40	4	4	4	-	1	1	-
> 40-50	1	-	-	-	1	1	-
> 50-100	6	5	4	-	2	-	1
> 100	4	4	2	-	1	-	-
all species	345	79	32	42	270	97	106

Average latitudinal and longitudinal range and average area size along a latitudinal gradient - Generally, the latitudinal and longitudinal ranges and the average area size increase considerably from South to North (Fig. 6D), it is the opposite of the species richness gradient (Fig. 6C). All three parameters have their lowest values at the latitudes of the Tropical Mexican Region, i.e. between 15° and 31-36°, but nevertheless increase considerably going from Guatemala and Southern Mexico to Northern Mexico and the Southern USA. In the temperate region, i.e. North of 31°, the increase in area size and latitudinal and longitudinal ranges is much more pronounced, especially for the latter parameter. The highest values are found in Northern Canada and exceed the lowest values for the Tropical Region about 14.5-fold (latitude), 20.5-fold (longitude) and 37-fold (area size).

In Guatemala and Southern Mexico, the longitudinal range is larger than the latitudinal range, as a result of the West-East orientation of the Neovolcanic Belt and the Southern Mexican mountains and highlands. In Northern Mexico North of the Neovolcanic Belt, in contrast, the latitudinal range exceeds the longitudinal range, since many species largely follow the North-South orientated Sierra Madre Occidental and Oriental. In the temperate region with its considerable West-East expansion, the longitudinal range again clearly exceeds the latitudinal range.

Species turnover along a latitudinal gradient - Beta diversity, defined as "extent of species replacement [] along environmental gradients" (Whittaker, 1972), is calculated here with the formula developed by Wilson & Shmida (1984), $\beta T = [g(H)+l(H)]/2a$, in which beta turnover (βT) is the species turnover or combined gain (g) and loss (l) of species along a gradient (H) standardized with the average sample richness (a), based on the above latitudinal ranges. The gain and loss of species was measured for intervals of 2 degrees of latitude along a Southward gradient (increasing temperature) from 70° to 14° Northern latitude. The sample richness, i.e. the species number at a given latitude, was calculated in intervals of 30 degree minutes and the average sample richness is therewith the average of four samples in each 2 degrees latitude interval.

Since the beta (species) turnover depends on the extent of latitudinal ranges and is therewith generally highest in regions with many local endemics, values increase going Southwards from 70° to 14° (Fig. 7A) and principally reflect the latitudinal gradient of species richness (Fig. 6C). The differences between temperate and tropical regions are more pronounced in beta turnover than in species richness. Values in the temperate region are comparatively low due to the sparse representation of local endemics, whereas especially in Southern Mexico the numerous local endemics (Table 5, 6) account for very high values.

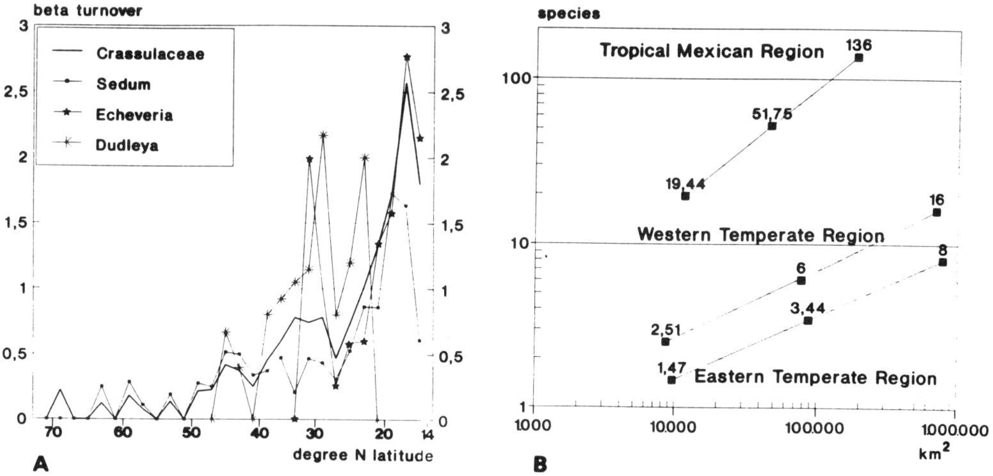

Figure 7. Beta (species) turnover and spatial floristic diversity of Northern American Crassulaceae. 7A. Beta (species) turnover along a latitudinal gradient calculated at intervals of 2 degrees of latitude (further explanations in the text). 7B. Estimations of spatial floristic diversity in the Eastern and Western Temperate and the Tropical Mexican Crassulaceae Regions based on the species inventar of the squares indicated in Figure 1 and 2 (further explanations in the text).

Differences in beta turnover in the three largest genera *Sedum, Echeveria,* and *Dudleya* reflect differences in latitudinal ranges. Temperate *Dudleya,* with latitudinal ranges comparable to tropical Crassulaceae (Table 5), also attains values for the beta turnover comparable to tropical Crassulaceae (Fig. 7A). Correspondingly, turnover values for *Sedum* are mainly below the average, since latitudinal ranges in *Sedum* are above average (Table 6). In contrast, *Echeveria* has the smallest average latitudinal range (Table 5), and consequently has the highest values for the beta turnover (Fig. 7A).

The maximal turnover values found between 18° and 16° latitude are 2.78 for *Echeveria* and 2.57 for all Crassulaceae, respectively, and correspond with a complete change of species composition over a distance of only c. 222 km. The total number of changes in species composition between 70° and 14° (= c. 6,212 km) equals 14.46.

Spatial floristic diversity - The floristic richness of a geographical region is a balance of its spatial floristic diversity or heterogeneity and its species abundance per unit area. Both factors can be computed on the basis of a correlation between the species number (S) and the area size (X) with an exponential function of the Arrhenius formula (Malyshev, 1991), $S = A X^z$, in which the constant A assesses the species number per unit area and the exponent z indicates the heterogeneity of a flora. The most accurate estimations of the constants A and z can be derived by least squares regression analysis, dividing a total area into a series of parts of different size (Malyshev, 1991). Such an analysis was undertaken in all three phytogeographical regions of Crassulaceae. In each of the low temperate regions a large quadrat (indicated in Fig. 1) encompassing 9 x 9 degree squares was successively divided into 9 quadrats of 3 x 3 degree squares and into 81 single degree squares. In the Tropical Mexican Region (Fig. 2) the quadrat encompassing 4 x 4 degree squares indicated was divided into 4 quadrats of 2 x 2 degree squares and into 16 single degree squares. For the analysis regions of highest species diversity were selected and degree squares with low species numbers were omitted as far as possible. The total species number was determined for all quadrats and averaged for each size class.

The corresponding graphs are almost exactly linear in a log-log plot (Fig. 7B) and enable the determination of the values for the constants A and z (Table 7). The value for the spatial floristic diversity (z) in the Tropical Mexican Region (0.702) represents the highest value reported up to now (Malyshev, 1991) and is considerably higher

Table 7. Estimates of spatial floristic diversity and floristic volume within the three phytogeographical Crassulaceae regions distinguished in Northern America in comparison with spatial floristic diversity of the surrounding total flora (the Tropical Mexican Crassulaceae Region is compared with the total flora of Central America, since the spatial floristic diversity of Mexico is unknown).

region	floristic volume (A)	spatial floristic diversity (z)	spatial floristic diversity of total flora (Malyshev, 1991)
Eastern Temperate	0.0428	0.386	0.189 (central region USA)
Western Temperate	0.0542	0.422	0.205 (mountain regions USA)
Tropical Mexican	0.0273	0.702	0.162 (central America)

than in both temperate regions, which exhibit similar values (0.386 and 0.422). However, these values may be influenced by incomplete knowledge of distribution ranges, especially in Mexico. In all three regions the spatial floristic diversity in Crassulaceae is much higher than that of the corresponding total flora (Table 7) and surpasses the latter more than twofold (temperate region) or even more than fivefold (Mexico).

Phytogeography, species richness, and evolution of Mexican Crassulaceae

Distribution of Crassulaceae in different Mexican vegetation types - The predominant distribution of Mexican Crassulaceae in mountain regions coincides with their abundance in pine-oak forests (Table 8), which cover major parts of the Mexican mountain ranges (Rzedowski, 1983). Crassulaceae are also somewhat more abundant in the higher, less arid matorral-desert scrub formations of the Southern Central Mexican Plateau. The arid matorral formations of the Chihuahuan and Sonoran Desert are almost completely devoid of Crassulaceae (Fig. 2). Few species occur in the tropical rain forests (5 *Echeveria* species), cloud forests (8 species), and the subalpine and alpine vegetation above the timberline (6 species), and only one species (*Thompsonella colliculosa* Moran) is reported from the tropical deciduous forests. In a general outline, tropical Mexican Crassulaceae occur preferentially in vegetation types with moderate temperatures and a not too low rainfall, which coincides well with their altitudinal zonation (Fig. 6B) and earlier descriptions for the habitat preferences of Mexican *Sedum* (Clausen, 1979).

Most lowland locations of tropical Mexican Crassulaceae are apparently situated in wet forests or ravines. One of the few species extending into pronounced arid lowlands is the xeromorphic *Echeveria laui* Moran & Meyrán, which occurs in arid tropical, succulent scrub vegetations at 500-550 m in the Tomellin-Canyon in Northern Oaxaca, but only on North-exposed and thus shaded rock faces (Lau, 1988). Some differences

Table 8. Number of species and percentages of Tropical Mexican Crassulaceae species in different vegetation types.

Vegetation zone	species number	%	*Echeveria* number	%	*Sedum* number	%
tropical deciduoud forest	1	1.0	-	-	-	-
tropical rain forest	5	4.8	5	9.8	-	-
matorral-desert shrub	20	19.0	12	23.5	5	11.6
pine-oak forest	64	61.0	29	56.9	31	72.1
cloud forest	8	7.6	5	9.8	2	4.7
subalpine-alpine vegetation	6	5.7	2	3.9	2	4.7
total	105		51		43	

between the relative abundance of *Sedum* and *Echeveria* are remarkable. *Echeveria* is relatively more abundant in matorral and less abundant in pine-oak forests than *Sedum*. The higher percentage of *Echeveria* in cloud and tropical rain forests may be due to the higher number of epiphytic species in *Echeveria* (21.6%, in contrast to 11.6% in *Sedum*).

Regional centres of species richness and endemism - Based on the distribution patterns of all species, eleven regional centres of species richness and endemism can be distinguished in Mexico and one in Guatemala. They represent phytochoria of restricted distribution with both a high total species number and a high number of local or regional endemics. These centres are delimited by the distribution areas of their endemic species (Fig. 8) and by and large correspond with the regions of higher species richness mentioned above. That these centres represent natural and well delimited phytogeographical units is clearly demonstrated by quantitative data. In a cluster analysis of all species, eleven of the twelve clusters contain by and large the species inventar of one of each of these centres (Chiapas is the single exception).

The considerable differences in the total number of species and number and percentage of endemics within these twelve regional centres reflect the differences in overall species richness (Fig. 2, Table 9). The percentage of endemic species varies twofold from 23.8% to 48.1% and is significantly correlated with the total species number (r=0.913 > 0.1%)

Table 9. Number of species, endemism, average degree of N latitude, and area index for the twelve regional centres of species richness and endemism of Figure 8.

no.	Regional centre of endemism	species	endemic species	endemic species %	degree N latitude	area index (104 km^2)
1	Barranca del Cobre	21	5	23.8	27° 30'	8.87
2	Mazatlán-Durango	17	5	29.4	23° 40'	8.03
3	Jalisco	25	10	40.0	20° 20'	7.49
4	Sierra Madre Oriental	30	12	40.0	25° 00'	7.23
8	Orizaba	35	13	37.1	19° 20'	7.16
5	San Luis Potosí	31	10	32.3	22° 10'	7.07
7	México-Morelos	47	16	34.0	19° 00'	6.98
11	Chiapas	13	5	38.5	16° 40'	6.40
6	Hidalgo-Querétaro	55	16	29.1	20° 40'	6.11
9	Tehuacán	48	19	39.6	17° 50'	5.10
10	Oaxaca	52	25	48.1	16° 40'	5.07
12	Guatemala	18	8	44.4	15° 10'	4.83

The Orizaba, Tehuacán and Oaxaca centres overlap slightly at their common borders, but most endemics are restricted to one of these centres. The Orizaba centre is a region with very high rainfall (see maps in García, 1965) and may include especially humid-adapted species, whose Southward migration is excluded by the arid Tehuacán Valley at its Southern border. The border between the Tehuacán and Oaxaca regional centres coincides approximately with the Southern extension of the Tehuacán Valley at Cuicatlan and the valleys around Oaxaca City, where more arid conditions may have prevented floristic exchange at least partially. This is best expressed in the ditypic *Sedum* sect. *Centripetalia*, with *S. allantoides* Rose endemic to the Tehuacán and *S. platyphyllum* Alexander endemic to the Oaxaca regional centre. The area of other species, e.g. *Echeveria purpusorum* Bgr., however, covers both regional centres.

Index of similarity - The floristic similarity at the species as well as the (infra)generic level between the twelve above regional centres was calculated with the index of similarity (IS) of Sörensen, IS (%) = (2C / A + B) * 100, in which C is the number of shared species between two regional centres, and A and B are the number of species or number of genera or infrageneric groups in each of the two regional centres A and B.

Table 10. Sörensen-indices of similarity on generic, infrageneric, and species level between the 12 regional centres of species richness and endemism of Figure 8. Values are emphasized in intervals of 10% (species) and 15% (generic - infrageneric level) by bold print and underlinings.

regional centre of species richness and endemism	species	genera/ infragen. groups	number	1	2	3	4	5	6	7	8	9	10	11	12
Bca. del Cobre	21	12	1		19,0	7,7		7,1	11,1	**32,3**	14,3		6,9		
Mazatlán-Durango	17	9	2	**21,1**		**43,5**	9,1	24,0	18,2	28,6	16,0	6,7	15,4		
Jalisco (Guadalajara)	25	14	3	4,3	**33,3**		7,4	**33,3**	**36,8**	**42,4**	**33,3**	11,4	19,4		7,4
Sierra Madre Oriental	32	13	4		4,1	3,5		**41,4**	**48,6**	12,5	13,8	5,9	13,3	10,0	
San Luis Potosí	31	16	5	3,8	16,7	**21,4**	**25,4**		**60,0**	28,6	18,9		6,1		
Hidalgo/Querétaro	55	24	6	5,3	8,3	20,0	**25,3**	**39,5**		**55,8**	**50,0**	**31,1**	24,4	19,4	10,8
México/Morelos	46	19	7	14,9	15,9	**25,4**	5,1	13,0	**29,7**		**57,1**	**40,0**	**38,9**	15,4	12,5
Orizaba	34	16	8	7,3	7,8	16,9	6,1	9,2	**27,0**	**32,5**		**54,1**	**48,5**	**34,8**	20,7
Tehuacán	49	21	9		3,0	5,4	2,5		15,4	**21,1**	**31,3**		**63,2**	28,6	23,5
Oaxaca	52	17	10	2,7	5,8	7,8	4,8	2,4	11,2	**22,4**	**25,6**	**41,6**		**33,3**	**33,3**
Chiapas	13	7	11				4,4		8,8	6,8	**21,3**	19,4	**21,5**		30,0
Guatemala	18	13	12			4,7			5,5	6,2	11,5	11,9	14,3	19,4	

index of similarity on generic / infrageneric level

index of similarity on species level

In most cases, the indices of similarity (Table 10) reach similar values at both taxonomic levels, but the indices at the (infra)generic level are normally higher than at the species level, since most (infra)generic groups are more widespread than the individual species. Higher indices are especially found between directly neighbouring regional centres and are therewith concentrated on both sides of the open diagonal. The maximum values at both levels are reached between the adjacent Tehuacán and Oaxaca regional centres. More expressed floristic similarities diverge from the core region of Crassulaceae (regional centres 6 to 10) along the major mountain ranges North-Westwards [Jalisco (3) - Mazatlan-Durango (2) - Barranca del Cobre (1)], Northwards

[San Luis Potosí (5) - Sierra Madre Oriental (4)], and South-Eastwards [Chiapas (11) - Guatemala (12)]. Zero values, on the other hand, are found especially between distant regional centres and are therewith concentrated at the corners of both triangles, though only 11 out of 66 comparisons (16.7%) have no species or (infra)generic taxon in common.

Chorograms for the regional centres of endemism and hypothetical routes of migration - A chorogram is defined as "a geographic map reporting a set of joint OGUs characters in a given OGS" and summarizes biogeographic data on a quantitative basis (Nimis & Bolognini, 1990). In this study, twelve chorograms (Fig. 9, 10) are shown which indicate the total distribution of the species occurring in each of the twelve regional centres outlined above and give a synthetic view of their phytogeographical affinities.

The major evolutionary centre for tropical Mexican Crassulaceae ranges nearly continuously from the states of Hidalgo and Querétaro to Oaxaca and includes the Hidalgo-Querétaro, México-Morelos, Orizaba, Tehuacán and Oaxaca regional centres (6-10 in Fig. 8), which are interconnected in many ways by expressed floristic similarities (Fig. 9F, 10A-D, Table 10) and comprise higher numbers of species and (infra)generic taxa, higher numbers and percentages of endemic species (also at the infrageneric level, Table 3, 9), and most of the centres of species richness within infrageneric groups of tropical *Sedum* and *Echeveria*. *Echeveria* series *Induplicatae* and *E.* series *Paniculatae*, which are basal within the genus and represent a link to Mexican *Sedum*, in particular to *S.* sect. *Pachysedum*, in their morphology, cytology, and crossing behaviour (Walther, 1972; Moran, 1978; Uhl, 1976a, 1994b), are also endemic to this region (Table 3).

From this major evolutionary centre, three hypothetical routes of migration lead North-Westwards, Northwards and South-Eastwards. These routes have been inferred from the chorograms of the more isolated regional centres (Fig. 9A-E, 10E-F) and reflect the floristic similarities along the Mexican mountain ranges (Table 10):

1) The chorograms of the Barranca del Cobre and the Mazatlan-Durango regional centres (Fig. 9A,D) show only very few floristic relations (Table 10). The latter centre is especially connected with the México-Morelos regional centre (Fig. 10B), but also shows more pronounced relations to West Central Mexico (Nueva Galicia). The Barranca del Cobre regional centre (Fig. 9A), in contrast, is again connected with the México-Morelos regional centre, but almost without any affinity with Nueva Galicia, and also has strong relations with Northern regions, Arizona and New Mexico, where the chorogram overlaps with the species of the Sierra Madre Oriental regional centre (Fig. 9B). These different floristic affinities might indicate that the Crassulaceae flora of both regional centres evolved largely independently from different sources. The Mazatlan-Durango centre, which has a higher rainfall, possibly came from Western Mexico along the humid Western Sierra Madre Occidental, whereas the Barranca del Cobre centre which has a more arid climate, evolved along the drier Eastern parts of the Sierra, especially from the region of the México-Morelos regional centre (see maps in García, 1965). Chromosome numbers of *Sedum* in the US-Mexican border region and North-Western Mexico are the lowest within Mexico (Uhl, 1985), which might imply

a pathway from North to South for this genus.

2) The Sierra Madre Oriental regional centre (Fig. 9B) shows similarities with the San Luis Potosí and Hidalgo-Querétaro regional centres (Fig. 9E, 10A), South-Western Texas (Big Bend region), Southern New Mexico and Southern Arizona, parts of the Chihuahuan Desert, and even the Sierra Madre Occidental, but it has almost no relations with Western Mexico and regions South of the Neovolcanic Belt. These affinities may indicate a migrational pathway from East Central Mexico along the Sierra Madre Oriental into the Southern USA. This migration route is best expressed in *Echeveria* series *Angulatae* and *Sedum* group *Moranense* (Table 3).

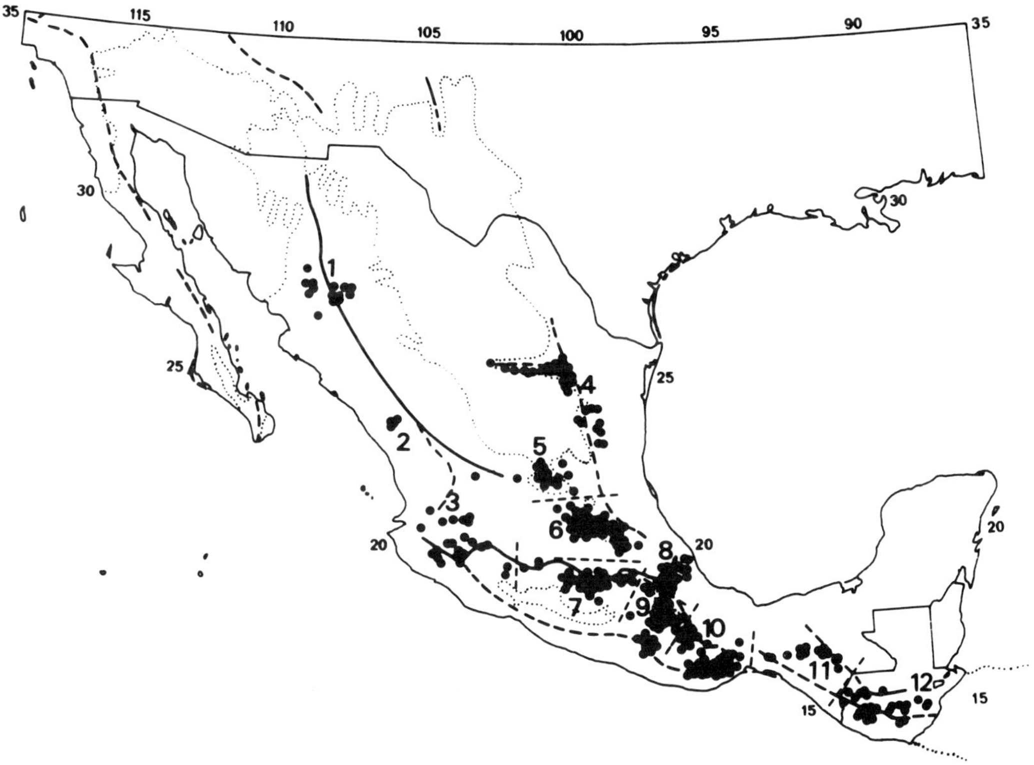

Figure 8. The Mexican and Guatemalan regional centres of species richness and endemism of the Crassulaceae flora as indicated by the distribution of their endemic species (each dot may represent more than one locality). 1. Barranca del Cobre. 2. Mazatlán-Durango highway. 3. Jalisco (Guadalajara). 4. Sierra Madre Oriental. 5. San Luis Potosí. 6. Hidalgo-Querétaro. 7. México-Morelos. 8. Orizaba. 9. Tehuacán. 10. Oaxaca. 11. Chiapas. 12. Guatemala.

3) The chorogram of the Guatemala regional centre (Fig. 10F) shows a clear restriction of most species to this centre and exhibits no expressed relations to Southern Mexico (Chiapas and Oaxaca). Chiapas (Fig. 10E), in contrast, shows strong affinities with Oaxaca-Puebla especially, but nearly none with Guatemala. This might again indicate that most probably the Crassulaceae floras of Chiapas and Guatemala developed largely independently from Southern Mexican lineages, almost without floristic exchange.

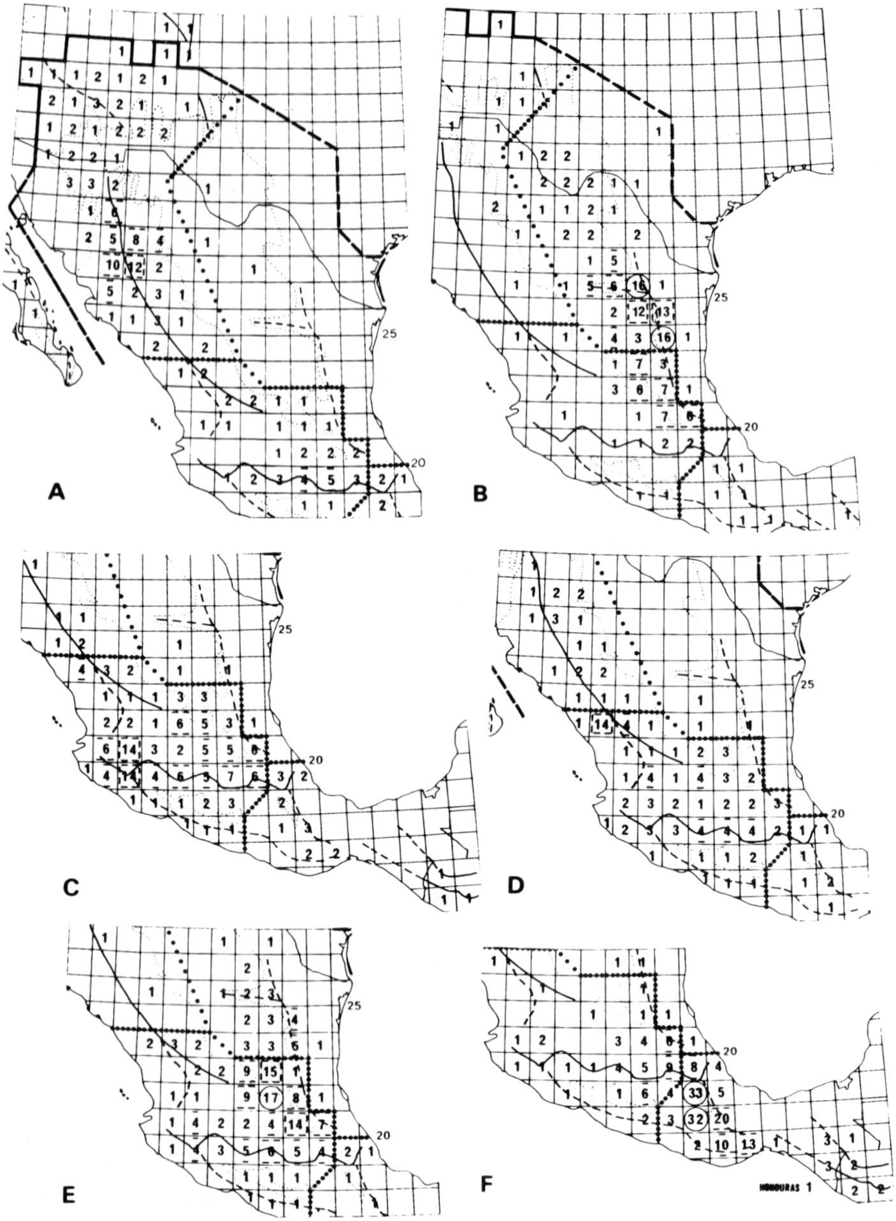

Figure 9. Chorograms of Mexican regional centres of species richness and endemism. 9A. Barranca de Cobre. 9B. Sierra Madre Oriental. 9C. Jalisco (Guadalajara). 9D. Mazatlán-Durango highway. 9E. San Luis Potosí. 9F. Tehuacán.

This migrational pathway finally leads via Central into South America, of which the regional centres principally represent species-poor Southern outliers of the Tropical Mexican Region.

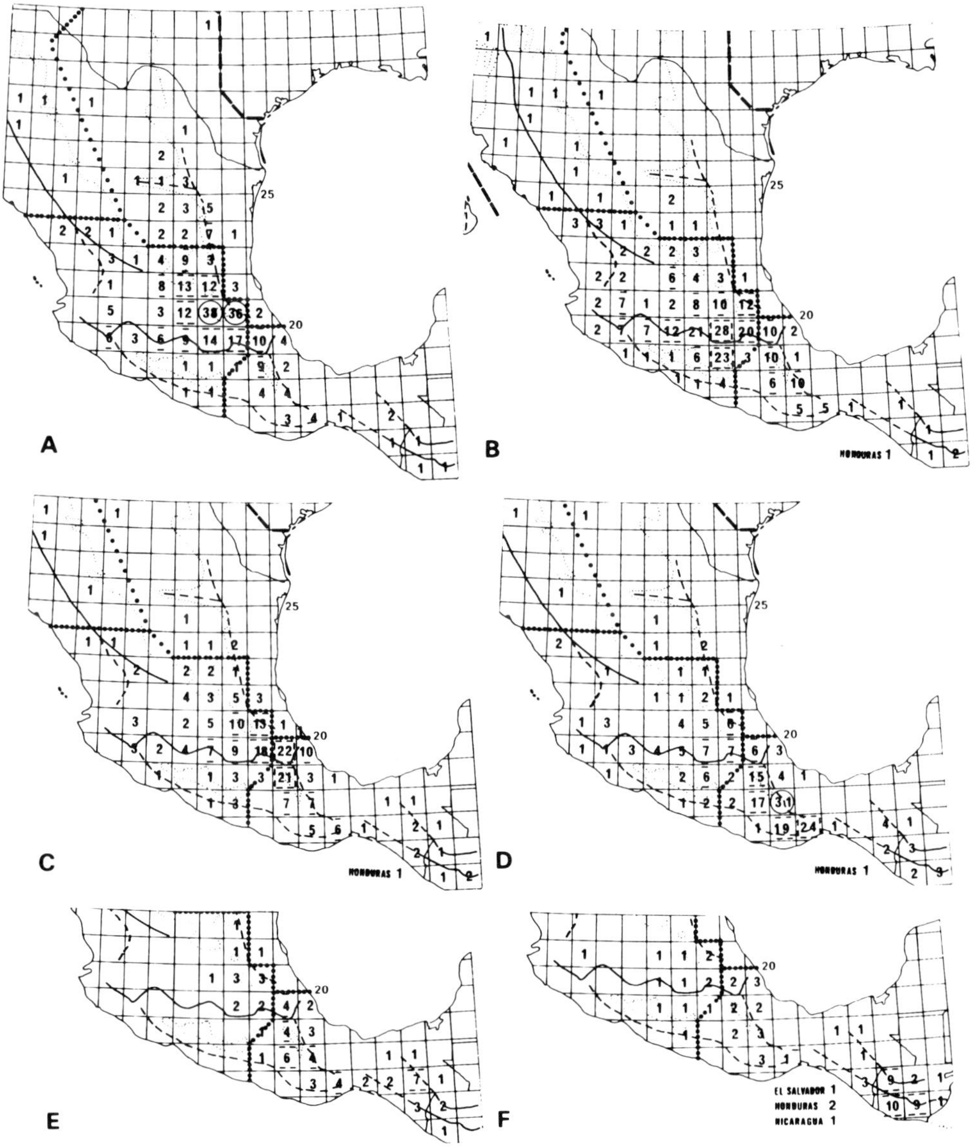

Figure 10. Chorograms of Mexican regional centres of species richness and endemism. 10A. Hidalgo-Querétaro. 10B. México-Morelos. 10C. Orizaba. 10D. Oaxaca. 10E. Chiapas. 10F. Guatemala.

The area index for the single chorograms - For the 12 chorograms, the "area index" (Nimis & Bolognini, 1990) was calculated. This gives a measure for the geographic extension of the species in each of the 12 chorograms. The area index is independent from the total species numbers and enables quantitative comparison of the chorograms. The area index values are shown in Table 9.

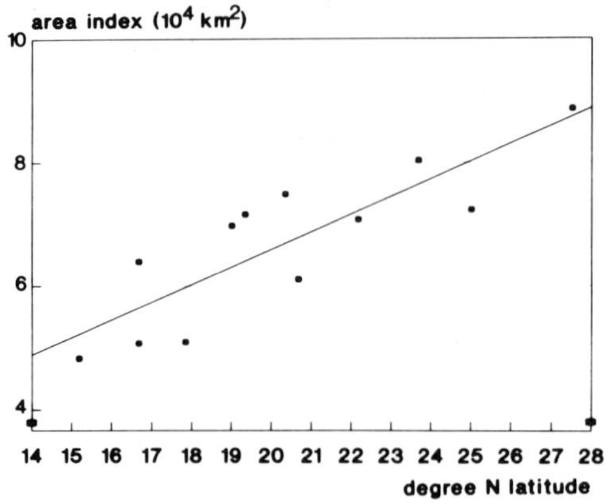

Figure 11. Correlation between the area index and the latitude of the twelve regional centres of species richness and endemism.

The area index of the chorograms varies about twofold, from the Guatemala Centre (Fig. 10F) with a very restricted areal extent, to the Barranca del Cobre Centre (Fig. 9A) with a pronounced area extensions to the North and the South.

Correlation of the mean latitude of the centres of endemism with the area index (Fig. 11, Table 9) reveals that area index values increase significantly ($r=0.7126 > 1\%$) from South to North. In other words, the relative amount of local endemics and species with a restricted distribution increases towards tropical, Southern latitudes in a flora (Fig. 6D, 7A, Table 5, 6), whereas the relative amount of more widespread species decreases concomitantly, and vice versa towards Northern latitudes. A further correlation of the area index with the number of species of the centres of endemism indicates decreasing area indices with increasing species number, but this correlation is statistically not significant ($r=-0.43075 > 5\%$).

South American Crassulaceae

Most probably Crassulaceae did not reach Central and South America before the emergence of a continuous Central American land bridge which was established only a few million years ago (Raven & Axelrod, 1974; Coney, 1982).

Guatemala comprises almost all Central American species (17 out of 18 species in 13 infrageneric taxa) and represents the Southernmost extension of the species-rich Mexican Crassulaceae flora (Table 2, 3). Central America South of Guatemala comprises only 5 species including one endemic (*Echeveria australis* Rose).

In South America the perennial Crassulaceae are restricted to the Andean mountains from Venezuela to Northern Argentina, whereas the annuals (*Tillaea*) extend into the lowlands East of the Andes (Table 2, 3; Bywater & Wickens, 1983). The highest species richness is found in Peru and Argentina (Table 2, 10).

South American *Echeveria* species belong to two series which have different distribution patterns (Table 3). The predominantly Mexican *E.* series *Nudae* is connected with species in Venezuela, Colombia and Ecuador across Central America. The continuous distribution and the possible conspecifity of the Venezuelan *E. bicolor* H.B.K. with the

Mexican-Guatemalan *E. montana* Rose (Moran, 1965) and the remarkable caryological uniformity of *Echeveria* in Venezuela (Kimnach & Uhl, 1992) may imply a relatively recent introduction. In contrast, the disjunct distribution of *E.* series *Racemosae* from Mexico and Guatemala to the Southern Andean states (Table 3) and especially the high chromosome numbers of the Bolivian taxa (Kimnach, 1991) suggest an early introduction and longer isolation. *Echeveria* may have traversed the humid Central American forests as an epiphyte. This growth-form is quite common in *E.* series *Nudae.*

The predominantly Mexican *Villadia* is absent from the region between Guatemala and Peru (Table 3), but has a secondary centre of diversity in the latter country.

There are no unambiguous records of *Sedum* species native to South America. The Bolivian *S. cymatopetalum* Fröd. and *S. jujuyense* Zardini from Argentina have basally connate petals (Zardini, 1971) and may belong to *Villadia.* The Peruvian *S. backebergii* Praeger is only known from the type collection. Furthermore, *S. praealtum* DC. is most probably not native South of Guatemala (Clausen, 1959) and *S. mexicanum* Britt. is an alleged introduction of Asian origin (Praeger, 1921; Uhl & Moran, 1972) .

General discussion of evolutionary patterns

The three Crassulaceae regions differ clearly in the number of species, genera and infrageneric taxa (Table 4), altitudinal zonation (Fig. 6D, 7A), average area size and average latitudinal and longitudinal range (Table 5, 6 and Fig. 6D), area index (Table 9), species turnover (Fig. 7A), spatial floristic diversity (Table 7), and cytologically, as the basic chromosome numbers of most tropical groups are higher than those of temperate groups (Uhl, 1985). However, in all parameters the two temperate regions are most similar and differ considerably from the Tropical Mexican Region. This points towards different patterns of evolutionary diversification in the temperate and tropical regions, which relate to the different climatical and ecological conditions.

Crassulaceae occur mainly in rock habitats, which are edaphically arid relative to the surrounding zonal vegetation, and thus represent azonal vegetation in the sense of Walter & Breckle (1983). Such habitats are described as typical for Mexican Crassulaceae (Uhl, 1985, 1992b, 1993) and also for South African and Macaronesian Crassulaceae (Jürgens, 1986; Thiede, 1991). This habitat preference leads to dispersed population structures with many small isolated populations and favours the evolution of local variants and local endemics. This pattern is expected to be much more pronounced in tropical and subtropical regions which often comprise extreme, small scale differences in topography, soil, and climate in comparison to temperate regions, where these parameters usually do not differ over large distances. The extraordinary high value of spatial floristic diversity in Southern Mexico (Table 7) reflects the varied climatical, topographical, and geological conditions of this region as well as its turbulent past marked by advances and retreats of glaciations and the accompanying climatic fluctuations and altitudinal and latitudinal shifts of vegetation belts (Lauer, 1986), and the emergence and uplift of mountains (Ferrusquia-Villafranca, 1993). Most probably a combination of these factors favours the evolution of large numbers of local endemics (Fig. 6D, 7A, Table 5, 6) and accounts for the enormous species richness of Crassulaceae in Southern Mexico (Fig. 2, 6C). The temperate region, in contrast, comprises mainly widespread species and only few local endemics (Fig. 6D, 7A, Table 5, 6).

Additionally, the rocky sites occupied by Crassulaceae are unstable pioneer habitats with a short life-span either as a result of competition with non-succulent plants after substrate weathering and accumulation or complete erosion of the whole location. In typical Mexican Crassulaceae habitats this leads to a marginal, shifting existence of populations continually settling in new habitats and dying out at old ones (Uhl, 1985, 1992b, 1993). *Dudleya pulverulenta* (Nutt.) Britt. & Rose may serve as an example. It is a pioneer on rock walls, road cuts, or disturbed soils, which is almost completely outcompeted by higher and non-succulent, faster growing, plants within ten to fifteen years (Thomson, 1993). Similar pioneer vegetations of Crassulaceae (*Sedum*, *Sempervivum*) in rock habitats in the Swiss Alps have also been described (Braun-Blanquet & Sutter, 1982). These short-term changes lead to highly dispersed and ephemeral population structures on a geographical as well as a time scale and fosters the evolution of new species of often restricted distribution. This process is probably favoured by the small seeds of Crassulaceae, which are easily dispersed by wind at least over short distances. This evolutionary model corresponds well with models of "disturbance ecology" which are regarded as key processes in the diversification of tropical biota (Godammer, 1992).

Acknowledgements

The author is especially indebted to Dr. H. 't Hart (Utrecht) for his kind permission to refer to his unpublished chloroplast DNA data of American Crassulaceae, and to T. Becker (Köln) for performing the cluster analysis. Dr. H. 't Hart, Prof. Dr. N. Jürgens (Köln), M. Kimnach (Reseda), Dr. J. Meyrán G. (Mexico City), Prof. Ch. Uhl (Ithaca), Prof. Dr. H. Uhlarz (Kiel), and B. Ullrich (Würm) are acknowledged for additional information, comments on the manuscript, and/or the supply of literature, and B. Friedrich (Kiel) for her help in drawing the figures. This publication is part of a dissertation project at the Faculty of Biology of the University of Kiel (Germany).

References

Bartel, J.A. 1993. *Dudleya*. In Hickman, J.C. (ed.), The Jepson Manual. Higher Plants of California, 525-530. University of California Press, Berkeley and Los Angeles.

Berger, A. 1930. Crassulaceae. In Engler, A. & Prantl, K. (eds.), Die natürlichen Pflanzenfamilien. ed. 2, **18a**, 352-483. Verlag Wilhelm Engelmann, Leipzig.

Böttcher, W. & Jäger, E.J. 1984. Zur Interpretation der Gattung *Sedum* L. s.l. (Crassulaceae) und ihrer Wuchsformtypen. *Wiss. Zeitschr. Univ. Halle* **33**: 127-141.

Braun-Blanquet, J. & Sutter, R. 1982. Zur Kenntnis der Crassulaceen-Pioniergesellschaft in den Bündner Alpen. Fragmenta Phytosociologica Raetica XII. *Jahresber. naturf. Ges. Graubünden* **99**: 75-83.

Brosius, G. 1989. SPSS/PC+ advanced statistics and tables. McGraw-Hill Book Company.
Bywater, M. & Wickens, G.E. 1983. New world species of the genus *Crassula*. *Kew Bull.* **39**: 699-728.
Cházaro Basañes, M.J. 1992. Phytogeographical aspects of the Crassulaceae of the Trans-Mexican Volcanic Belt. *IOS Bull.* **5**(4a): 134-135.
Clausen, R.T. 1959. *Sedum* of the Trans-Mexican Volcanic Belt: an exposition of taxonomic methods. Cornell University Press, Ithaca.
Clausen, R.T. 1975. *Sedum* of North America north of the Mexican Plateau. Cornell University Press, Ithaca, New York.
Clausen, R.T. 1977. Biennal species of *Sedum* of the Sierra Madre Occidental and Mexican Plateau. *Bull. Torrey Bot. Club* **104**: 209-217.
Clausen, R.T. 1978. *Sedum* - seven Mexican perennial species. *Bull. Torrey Bot. Club* **105**: 214-223.
Clausen, R.T. 1979. *Sedum* in six areas of the Mexican Cordilleran Plateau. *Bull. Torrey Bot. Club* **106**: 205-216.
Clausen, R.T. 1981. Variation of species of *Sedum* of the Mexican Cordilleran Plateau. Arnold Printing Corp., Ithaca, New York.
Clausen, R.T. 1984. *Sedum* (Crassulaceae) of the Mexican Cordilleran Plateau. *Gentes Herb.* **12**: 8-48.
Coney, P.J. 1982. Plate tectonic constraints on the biogeography of middle America and the Caribbean region. *Ann. Missouri Bot. Gard.* **69**: 432-443.
Cox, C.B. & Moore, P.D. 1993. Biogeography. An ecological and evolutionary approach. Blackwell Scientific Publ., Oxford.
Crovello, T.J. 1981. Quantitative phytogeography: an overview. *Taxon* **30**: 563-575.
Delgadillo M., C. 1971. Phytogeographic studies on alpine mosses of Mexico. *Bryologist* **74**: 331-346.
Delgadillo M., C. 1987. Moss distribution and the phytogeographical significance of the Neovolcanic Belt of Mexico. *J. Biogeogr.* **14**: 69-78.
Denton, M.F. 1982. Revision of *Sedum* section *Gormania* (Crassulaceae). *Brittonia* **34**: 48-77.
Denton, M.F. 1993. *Parvisedum, Sedum*. In Hickman, J.C. (ed.), The Jepson Manual. Higher Plants of California, 531-534. University of California Press, Berkeley and Los Angeles.
Devender, T.R. van & Spaulding, W.G. 1979. Development of vegetation and climate in the southwestern United States. *Science* **204**: 701-710.
Ferrusquia-Villafranca, I. 1993. Geology of Mexico: a synopsis. In Ramamoorthy, T.P., Bye, R., Lot, A. & Fa, J. (eds.), Biological diversity of Mexico: origins and distribution, 3-108. Oxford University Press, New York, Oxford.
Fröderström, H. (1935): The genus *Sedum* L. A systematic essay. 4. *Acta Horti Gothob.* **10**: 2-262.
García, E. 1965. Distribución de la precipitación en la República Mexicana. *Publ. Inst. Geogr. Méx.* **1**: 173-191.
Goldammer, J.G. (ed.) 1992. Tropical forests in transition: ecology of natural and anthropogenic disturbance processes. Birkhäuser Verlag, Basel, Boston, Berlin.
Ham, R.C.H.J. van. 1994. Phylogenetic implications of chloroplast DNA variation in the Crassulaceae. Ph.D. Thesis, Utrecht University.
Ham, R.C.H.J. van, Hart, H. 't, Mes, T.H.M. & Sandbrink, J.H. 1994. Molecular evolution of noncoding regions of the chloroplast genome in the Crassulaceae and related species. *Curr. Genet.* **25**: 558-566.
Hart, H. 't. 1991. Evolution and classification of the European *Sedum* species (Crassulaceae). *Fl. Medit.* **1**: 31-61.
Hernández, H. 1994. The geographical distribution of Mexican endangered cacti, with special ref-

erence to the Chihuahuan Desert species. *IOS Bull.* **5**(5): 253.

Jürgens, N. 1986. Untersuchungen zur Ökologie sukkulenter Pflanzen des südlichen Afrika. *Mitt. Inst. Allg. Bot. Hamburg* **21**: 139-365.

Jürgens, N. 1991. A new approach to the Namib region. I: Phytogeographic subdivision. *Vegetatio* **97**: 21-38.

Kimnach, M. 1991. *Echeveria bakeri*, a new species from Bolivia. *Cact. Succ. J. (US)* **63**: 254-257.

Kimnach, M. & Uhl, C.H. 1992. Notes on *Echeveria* in Venezuela. III: *Echeveria pendulosa*, sp. nov. *Cact. Succ. J. (US)* **64**: 200-203.

Knapp, U. 1994. Skulptur der Samenschale und Gliederung der Crassulaceae. *Bot. Jahrb. Syst.* **116**: 157-187.

Lau, A.B. 1988. Weitere Erkundigungen über *Echeveria laui*. *Mittelstachel* (Würzburg) **7**: 132-137.

Lauer, W. 1986. Die Vegetationszonierung der Neotropis und ihr Wandel seit der Eiszeit. *Ber. Deutsch. Bot. Ges.* **99**: 211-235.

Lauer, W. 1993. Klimatologie. Westermann Schulbuchverlag, Braunschweig.

MacMahon, J.A. & Wagner, F.H. 1985. The Mojave, Sonoran and Chihuahuan Deserts of North America. In Evenari, M., Noy-Meir, I. & Goodall, D.W. (eds.), Hot deserts and arid shrublands, A. Ecosystems of the world **12A**, 105-202. Elsevier, Amsterdam.

Malyshev, L. 1991. Some quantitative approaches to problems of comparative floristics. In Nimis, P.L. & Crovello, T.J. (eds.), Quantitative approaches to phytogeography, 15-34. Kluwer Academic Publishers, Dordrecht.

Malyshev, L., Nimis, P.L. & Bolognini, G. 1994. Essays on the modelling of spatial floristic diversity in Europe: British Isles, West Germany, and East Europe. *Flora* **189**: 79-88.

McDonald, J.A. 1993. Phytogeography and history of the alpine-subalpine flora of Northeastern Mexico. In Ramamoorthy, T.P., Bye, R., Lot, A. & Fa, J. (eds.), Biological diversity of Mexico: origins and distribution, 681-706. Oxford University Press, New York, Oxford.

McLaughlin, S.P. 1989. Natural floristic areas of the western United States. *J. Biogeography* **16**: 239-248.

Moran, R. 1942. The status of *Dudleya* and *Stylophyllum*. *Desert Pl. Life* **14**: 149-157.

Moran, R. 1965. *Echeveria montana* Rose. *Cact. Succ. J. (US)* **37**: 178-183.

Moran, R. 1978. Resurrection of *Cremnophila* (Crassulaceae). *Cact. Succ. J. (US)* **50**: 139-146.

Moran, R. 1981. *Dudleya*. In Jacobsen, H. (ed.), Das Sukkulentenlexikon, ed. 2, 167-172. Gustav Fischer Verlag, Jena.

Moran, R. 1994. The genus *Thompsonella* Rose (Crassulaceae). *Haseltonia* **2**: 1-19.

Nimis, P.L. & Bolognini, G. 1990. The use of chorograms in quantitative phytogeography and in phytosociological syntaxonomy. *Fitosociologia* (Pavia) **25**: 67-87.

Ohba, H. 1978. Generic and infrageneric classification of the Old World Sedoideae (Crassulaceae). *J. Fac. Sci. Univ. Tokyo, Sect. 3, Bot.* **12**: 139-198.

Ohba, H. 1989. Biogeography of the genus *Rhodiola* (Crassulaceae), with special reference to the floristic interaction between the Himalaya and the arctic region. In Ohba, H., Hayami, I. & Mochizuki, K. (eds.), Current aspects of biogeography in West Pacific and East Asian regions. Nature & Culture, **1**, 115-133. University of Tokio, University Museum, Tokyo.

Pianka, E.R. 1966. Latitudinal gradients of species diversity: a review of concepts. *Amer. Natural.* **100**: 33-46.

Praeger, R.L. 1921. An account of the genus *Sedum*. Royal Horticultural Society, London.

Raven, P.H. & Axelrod, D.I. 1974. Angiosperm biogeography and past continental movements. *Ann. Missouri Bot. Gard.* **61**: 539-673.

Raven, P.H. & Axelrod, D.I. 1978. Origin and relationships of the California flora. *Univ. Calif. Publ. Bot.* **72**: 1-134.

Rzedowski, J. 1983. Vegetación de Mexico. Editorial Limusa, Mexico.

Rzedowski, J. 1993. Diversity and origins of the phanerogamic flora of Mexico. In Ramamoorthy, T.P., Bye, R., Lot, A. & Fa, J. (eds.), Biological diversity of Mexico: origins and distribution, 129-146. Oxford University Press, New York, Oxford.

Styles, B.T. 1993. Genus *Pinus*: a Mexican purview. In Ramamoorthy, T.P., Bye, R., Lot, A. & Fa, J. (eds.), Biological diversity of Mexico: origins and distribution, 397-420. Oxford University Press, New York, Oxford.

Takhtajan, A. 1986. Floristic regions of the world. University of California Press, Berkeley, Los Angeles.

Teeri, J.A., Stowe, L.G. & Murawski, D.A. 1978. The climatology of two succulent plant families: Cactaceae and Crassulaceae. *Can. J. Bot.* **56**: 1750-1778.

Thiede, J. 1991. Untersuchungen zur vegetativen Anatomie makaronesischer Sempervivoideae (Crassulaceae). Ms. Thesis, University of Hamburg.

Thomson, P. 1993. The *Dudleya* and *Hasseanthus* handbook. Bonsall Publications, California.

Tölken, H.R. 1985. Crassulaceae. In Leistner, O.A. (ed.), Flora of Southern Africa, 14. Dep. Agr. Wat. Suppl., Pretoria.

Uhl, C.H. 1970. Chromosomes of *Graptopetalum* and *Thompsonella* (Crassulaceae). *Amer. J. Bot.* **57**: 1115-1121.

Uhl, C.H. 1976a. Chromosomes, hybrids and ploidy of *Sedum cremnophila* and *Echeveria linguifolia* (Crassulaceae). *Amer. J. Bot.* **63**: 806-820.

Uhl, C.H. 1976b. Chromosomes of Mexican *Sedum* I. Annual and biennal species. *Rhodora* **78**: 629-640.

Uhl, C.H. 1978. Chromosomes of Mexican *Sedum* II. Section *Pachysedum*. *Rhodora* **80**: 491-512.

Uhl, C.H. 1980. Chromosomes of Mexican *Sedum* III. Sections *Centripetalia*, *Fruticisedum*, and other woody species. *Rhodora* **82**: 377-402.

Uhl, C.H. 1983. Chromosomes of Mexican *Sedum* IV. Heteroploidy in *Sedum moranense*. *Rhodora* **85**: 243-252.

Uhl, C.H. 1985. Chromosomes of Mexican *Sedum* V. Section *Sedum* and Subgenus *Sulcus*. *Rhodora* **87**: 381-423.

Uhl, C.H. 1992a Chromosomes of Mexican *Sedum* VI. Section *Sedastrum*. *Rhodora* **94**: 362-370.

Uhl, C.H. 1992b Polyploidy, dysploidy, and chromosome pairing in *Echeveria* (Crassulaceae) and its hybrids. *Amer. J. Bot.* **79**: 556-566.

Uhl, C.H. 1993. *Tacitus* and polyploidy. *Haseltonia* **1**: 29-33.

Uhl, C.H. 1994a. Intergeneric hybrids in the Mexican Crassulaceae: II. *Dudleya* (and plate tectonics). *Cact. Succ. J. (US)* **66**: 74-80.

Uhl, C.H. 1994b. Chromosomes and hybrids of *Echeveria* (Crassulaceae). I. Series *Induplicatae* and *Paniculatae*. *Haseltonia* **2**: 79-88.

Uhl, C.H. & Moran, R. 1972. Chromosomes of Crassulaceae from Japan and South Korea. *Cytologia* **37**: 59-81.

Uhl, C.H. & Moran, R. 1973. The chromosomes of *Pachyphytum* (Crassulaceae). *Amer. J. Bot.* **60**: 648-656.

Walter, H. & Breckle, S.W. 1983. Ökologie der Erde.**1**. Ökologische Grundlagen in globaler Sicht. Gustav Fischer Verlag, Stuttgart.

Walther, E. 1936. Phylogeny of *Echeveria*. *Cact. Succ. J. (US)* **8**: 82-88.

Walther, E. 1972. *Echeveria*. California Academy of Sciences, San Francisco.

Whittaker, R.H. 1972. Evolution and measurement of species diversity. *Taxon* **21**: 213-251.

Wilson, M.V. & Shmida, A. 1984. Measuring beta diversity with presence-absence data. *J. Ecol.* **72**: 1055-1064.

Zardini, E.M. 1971. Las especies del genero *Sedum* (Crassulaceae) espontaneas en la Republica Argentina. *Bol. Soc. Argent. Bot.* **14**: 95-106.

CHAPTER 7

Floristic and phytogeographical studies on the Crassulaceae of Jalisco (Mexico)

Miguel J. Cházaro Basañez & Joachim Thiede

Abstract

Since little or no floristic and phytogeographical research has been done on the Crassulaceae of Western-Central Mexico, we have started a study of this family in the state of Jalisco and the adjacent states about three years ago. So far 33 species of four genera of Crassulaceae and one hybrid have been found in Jalisco, i.e. *Echeveria agavoides, E. colorata, E. chapalensis, E. dactylifera, E. elegans, E. fulgens, E. lozanii, E. mucronata, E. pringlei, E.* x *sayulensis, Graptopetalum amethystinum, G. fruticosum, G. pachyphyllum, G. pentandrum* ssp. *superbum, Sedum bourgaei, S. ebracteatum, S. fuscum, S. grandipetalum, S. greggii, S. griseum, S. guadalajaranum, S. hintoni, S. jaliscanum, S. longipes, S. machucana, S. moranense, S. multiflorum, S. pringlei, S. semiteres, S. tortuosum, S. vinicolor, Villadia platystyla, V. batesii,* and *V. painteri.* The floristic and phytogeographic data clearly indicate that the major part of the Crassulaceae flora of the state of Jalisco is a combination of elements from the Crassulaceae floras of the Southern Sierra Madre Occidental, the Trans-Mexican Volcanic Belt and Southern part of the Central-Mexican Plateau, as well as the Sierra Madre del Sur. In addition it comprises a strong endemic element (c. 30%), which fully justifies future, more extensive studies of this rich flora.

Introduction

The senior author first became interested in the Crassulaceae in 1989, and, being more familiar with the flora of Eastern Mexico, started with a study of the taxa of the Central part of the state of Veracruz and adjacent Puebla (Cházaro, 1990a, 1990b). In

April 1991, together with C. Viveros, he rediscovered *Sedum nussbaumerianum* Bitter by mere accident in a ravine South of Xalapa in the state of Veracruz (Cházaro & Viveros, 1991) and this fostered his interest on the subject. At the same time he became aware of the lack of knowledge of the Crassulaceae of the state of Jalisco, and felt obliged to undertake a study of the stonecrop family for this region. In this paper a report on the results of these studies is presented to which the junior author has added some comments on the phytogeography of the Crassulaceae of Jalisco.

Material and Methods

For this study the traditional approach has been used, viz. a combination of literature studies, field work, and the revision of herbarium material. Since 1990 the senior author made several excursions in Jalisco and the adjacent states of Colima, Nayarit, Zacatecas, Durango, Aguascalientes, Guanajuato, and Michoacán and studied the Crassulaceae collections in the following herbaria (abbreviations, Index Herbariorum 1, ed. 8): CHAPA, ENCB, GUADA, IEB, MEXU, MICH, MO, WIS, and XAL.

Results and Discussion

Geographical features of Jalisco - The state of Jalisco is located in the Western-Central part of the Mexican Republic. It lies between 19° and 22° Northern latitude and 101°30' and 105°80' Western longitude. With 80,130 square kilometres it is the sixth largest state in the country. Six physiographic regions are distinguished within the state: 1) The Southern part of the Sierra Madre Occidental (or Sierra de los Huicholes); 2) The Central High Plateau (or Los Altos de Jalisco); 3) The Trans-Mexican Volcanic Belt (Nevado de Colima with an altitude of 4370 m being the highest point in Jalisco); 4) The Northern part of the Sierra Madre del Sur; 5) The coastal plains of the Pacific Ocean; 6) A small piece in the Balsas river basin (Jilotlan de los Dolores and Manuel M. Dieguez counties).

There is a trend of decreasing humidity from the South to the North. Rainfall at Sierra Manantlan and Sierra Cuale is approximately 1500 mm per year, whereas Guadalajara in the Central part of Jalisco receives 800-900 mm per year, and the precipitation in the inland village of Ojuelos is only about 400 mm per year.

Because of the differences in altitude, climate and soil there are several types of vegetation, viz. mangrove forest, palm stands, tropical semi-deciduous forest, savanna, tropical deciduous forest, oak forest, mesophytic forest, pine-oak-madroño forest, fir forest, pine forest, pinyon pine-juniper forest and xeric shrubland.

The Crassulaceae flora of Jalisco - The exploration of Crassulaceae flora of the state of Jalisco started with the collections by C.G. Pringle, E. Palmer, Y. Mexia, and J.N. Rose in the late 19th and early 20th century, and the species described from these specimens by S. Watson and J.N. Rose. More recently (in the 1950's, 60's and 70's) explorations by R. McVaugh, J. Rzedowski, H. Moore, R.T. Clausen, C.H. Uhl, M. Kimnach, J. Meyrán, R. Moran, and F. Otero resulted in a few new species as well as several new records for *Echeveria*, *Graptopetalum*, *Sedum*, and *Villadia*. However, our revision of herbarium collections revealed that the Crassulaceae are usually underrepresented, which may partly be due to their preference for inaccessible, rocky habitats.

For Jalisco we have so far (1994) discovered one new species, *Sedum machucana* Cházaro & Metzger (in prep.), two new records, viz. *Graptopetalum pentandrum* and *Sedum longipes*, and a large number of new localities for several species. 33 species of 4 genera of Crassulaceae occur in Jalisco, i.e. *Echeveria* with 9 species (and 1 hybrid), *Graptopetalum* with 4 species, *Sedum* with 17 species, and *Villadia* with 3 species.

Echeveria DC.

In 1958 E. Walther, the authority on *Echeveria*, knew only of the presence of two species of *Echeveria* in Jalisco, i.e. *E. lozanii* and *E. pringlei*. Gold (1968) reported the same two species for Jalisco, and added *E. tenuis* Rose. However, the latter species does not occur in Jalisco, as it is only known from the type locality at Monte Escobedo in the state of Zacatecas (Britton & Rose, 1905; Walther, 1972). Walther (1972), in a monograph on *Echeveria*, listed five species and the alleged hybrid *E.* x *sayulensis* for Jalisco. Presently the occurrence of nine species of *Echeveria* has been established and this brief account may serve as an example of the poor knowledge of this genus up till recently.

1. *E. agavoides* Lemaire. This easily recognizable species was reported by Reppenhagen (1985) from the flatlands and rocky hills of Ojuelos de Jalisco. *E. agavoides* has also been reported from the states of Guanjuato, Hidalgo, and San Luis Potosí.

2. *E. colorata* E. Walther. The species was originally described by Walther (1972) from plants at the University of California botanical garden. Kimnach (1980) reported two wild populations from Jalisco. He described *E. colorata* var. *brandtii* from the Sierra Tapalpa and reported var. *colorata* from the Volcan de Tequila. Cházaro & Machuca (1992) also gave these two locations and added one more for var. *colorata*, namely Cerro Viejo. After 1992 J.A. Machuca has found two more populations (Machuca, unpublished data), one at the Sierra Verde and one at Cerro Garcia, respectively. *E. colorata* is endemic to the volcanic mountains of Central Jalisco. It flowers in March, April and May. Because of the beauty of its red foliage it is sold at downtown Guadalajara city during the Christmas season.

3. *E. chapalensis* Moran & Uhl. This species was discovered by C.H. Uhl on the mountains West of Lake Chapala (Michoacán) in 1972, and described 17 years later (Moran & Uhl, 1989). It was reported by Cházaro & Machuca (1992) from the foothills of the Cerro Viejo volcano, which is only 10 miles NW of Lake Chapala. Huerta (1993) reported it from the Barranca del Salko de Nogal at Sierra Tapalpa. It is in bloom from March to April. Except for a few populations in the very Northern part of Michoacán, *E. chapalensis* is restricted to Jalisco.

4. *E. dactylifera* E. Walther. A large (to 1 m or more), perennial herb described from cultivated plants originating from Espinazo del Diablo (Western Sierra Madre), along

the road from Durango to Mazatlán. J.N. Rose collected it in Jalisco, on the Sierra de los Huicholes, West of Bolaños. Recently we found a large population of *E. dactylifera* at Cerro Colli, on the Western outskirts of the Guadalajara metropolitan area. In flower during September and October. *E. dactylifera* also occurs in the states of Aguascalientes, Durango, Sinaloa, and Michoacán (probably).

5. *E. elegans* Rose var. *tuxpanensis* E. Walther. This variety was described by Walther (1972) from plants collected in the Beltran cañon, near Tuxpan (Jalisco). So far it is only known from the type locality. The two other varieties of *E. elegans*, var. *elegans* and var. *hernandonis* E. Walther, respectively, are both restricted to the state of Hidalgo.

6. *E. fulgens* Lem. Known in Jalisco from the Volcan del Fuego, Nevado de Colima, Cerro Grande, and Apango (Sierra de Tapalpa). It flowers from October to December. This species is also known from the states of Durango, Guerrero, México, Michoacán, and Oaxaca.

7. *E. lozanii* Rose. A very poorly known species, which has never been found again after the initial collection made by C.G. Pringle and F. Lozano in 1903. The type locality is situated in the mountains West of Etzatlan (Jalisco). We have explored this area three times without success.

8. *E. mucronata* Schlechtendal. In Jalisco found at Las Piedrotas near Tapalpa. *E. mucronata* (including *E. platyphylla* Rose and *E. crassicaulis* E. Walther) also occurs in the states of Chihuahua, Distrito Federal, Hidalgo, Durango, Guanajuato, México, Michoacán, Oaxaca, Puebla, Tamaulipas, Veracruz, and Zacatecas.

9. *E. pringlei* (S. Watson) Rose. Described from plants collected by C.G. Pringle on a dry ledge in the Barranca de Guadalajara (Jalisco) in November 1888. The species is endemic to the type area in Jalisco. Flores & Cházaro (1993) reported some new localities along the big ravine near Guadalajara, i.e. at the Barranca de Tonala, the Barranca de Colimilla, the Barranca de Oblatos, and the Barranca de la Experiencia. *E. pringlei* flowers in November and December.

10. *E.* x *sayulensis* E. Walther. According to Uhl (1989) this taxon is a hybrid between *E. secunda* Booth. and *E. gibbiflora* DC. It was described in 1972 from cultivated material originating from Sayula (Jalisco).

Graptopetalum Rose

An almost exclusively Mexican genus. Four of the ten known species occur in Jalisco, viz. two in the Southern part and two in the Northern part.

1. *G. amethystinum* (Rose) E. Walther. In September 1897 J.N. Rose found a new Crassulaceae species at the top of the high Sierra Madre, West of Bolaños (Jalisco), which he described as *Pachyphytum amethystinum*. Later Walther transferred it to *Graptopetalum*. In September 1959 R. Moran and M. Kimnach found it on the mountains between Durango and Mazatlán (Moran, 1963). Like the Bolaños site, this second location of *G. amethystinum* is situated on the Western Sierra Madre. Ninety-three years after its discovery A. Flores collected *G. amethystinum* for the second time at or near the type locality West of Bolaños in June 1990 (Flores & Cházaro, in press). It flowers in June and July. *G. amethystinum* also occurs in the states of Durango, Hi-

dalgo, and Sinaloa.

2. *G. fruticosum* Moran. This species was described by Moran & Uhl (1968) from plants collected at Puerto Los Mazos, S of Autlán (Jalisco) by C.H. Uhl in April 1967. In addition Lomelí (1988) reported three locations in Southern Jalisco. Cházaro & Machuca (1992) reported it from Cerro Viejo, where it is very abundant in shady and humid ravines. *G. fruticosum* is endemic to the Southern and Central part of Jalisco. It flowers in March and April.

3. *G. pachyphyllum* Rose. A small and elusive plant, from the high plateau in Querétaro and Hidalgo. There are two records for Jalisco. The first is a collection by R. McVaugh from Chinampas (Ojuelos county) made in 1958, the second is by C. Glass from the road to Ojuelos at 32 km N of Lagos de Moreno made in November 1993. We have visited both localities, but have not been able to locate this species. It flowers in August. *G. pachyphyllum* also occurs in the states of Guanajuato, Hidalgo, Querétaro, San Luis Potosí, and Zacatecas.

4. *G. pentandrum* Moran ssp. *superbum* Kimnach. The species was described by Moran (1971) from material obtained by M. Kimnach in 1970 at a small nursery in Guadalajara (Jalisco). The origin of the plants was unknown. Kimnach (1987) described *G. pentandrum* ssp. *superbum* from a plant purchased by L. Aviña at La Barca (Jalisco), but this was also of unknown wild origin. Cházaro & Flores (1992) reported a natural population of ssp. *superbum* for the first time. They found the plants at the Barranca de los Corrales in April 1990. *G. pentandrum* ssp. *superbum* flowers in April and May. A. Lau found *G. pentandrum* ssp. *pentandrum* in the wild on the Sierra Coalcoman (Michoacán) in May 1973.

Sedum L.

The species richness of *Sedum* in Jalisco proved to be largely unknown up till now. Britton & Rose (1905) reported three species for Jalisco, viz. *S. guadalajaranum*, *S. jaliscanum*, and *S. moranense* (= *S. submontanum*). In 1959 Clausen mentioned only four species, viz. *S. ebracteatum*, *S. grandipetalum*, *S. jaliscanum*, and *S. tortuosum*, but in the following years raised the number to seven, adding *S. guadalajaranum*, *S. greggii*, and *S. multiflorum* to his list (Clausen, 1959, 1978, 1984). On the other hand, Gold (1968) in a survey of the succulents of Jalisco did not mention a single *Sedum* species.

1. *S. bourgaei* Hemsley. This species was collected by M. Kimnach and Boutin at about 2600 m between Laguna de Sayula and Tapalpa (Jalisco) in a region dominated by pines and oaks (Kimnach, 1980). We found it at Cerro Viejo and the adjacent regions (Cházaro & al., 1992). *S. bourgaei* also occurs in the states of Distrito Federal, México, Michoacán (probably), Morelos, Oaxaca, Querétaro, and Tlaxcala.

2. *S. ebracteatum* Sessé & Mociño ex DC. This herbaceous plant grows on rocky cliffs. The aerial parts wither during the dry season. It was first reported for Jalisco as *S. chapalense*, which was described by S. Watson from Chapala in 1886 (see Clausen, 1959). We have found it at the Oblatos ravine and on the mountains around lake Chapala. It produces its white flowers from October to December. *S. ebracteatum* also occurs in the states of Distrito Federal, Durango, Guanajuato, Guerrero, Hidalgo,

México, Michoacán, Oaxaca, Puebla, Querétaro, San Luis Potosí, Tamaulipas, and Zacatecas.

3. *S. fuscum* Hemsley. Not yet found in Jalisco, but very likely to occur in this state, since it was collected by R. McVaugh and J. Rzedowski at the Sierra de Laurel in the state of Aguascalientes not far from the Jaliscan border. *S. fuscum* has also been reported for the states of Durango and San Luis Potosí.

4. *S. grandipetalum* Fröderström. This species looks like a robust version of *S. greggii*. It has yellow flowers which appear from December to May, and grows on rocky cliffs. *S. grandipetalum* was described from a specimen collected by I. Mexia at Real Alto, Cerro La Bufa, San Sebastian del Oeste (Jalisco) in February 1927 (see Clausen, 1959). It is endemic to Southern Jalisco and is also known from the Sierra de Manantlán, the Nevado de Colima. Recently we collected it on the Volcan del Fuego.

5. *S. greggii* Hemsley. A herb with pendent branches and yellow flowers which appear from November to March. It is known in Jalisco from the Volcan de Tequila, the Sierra Tapalpa, and the Cerro Viejo volcano. Clausen (1978) described *S. greggii* ssp. *angustifolium*, which is endemic to Jalisco, from the Volcan de Tequila. *S. greggii* ssp. *greggii* also occurs in the states of Distrito Federal, Guanajuato, Hidalgo, México, Michoacán, Querétaro, San Luis Potosí, and Tamaulipas (probably).

6. *S. griseum* Praeger. Herbaceous colonizing plant, obviously very closely related to *S. bourgaei*, though it is generally somewhat larger with thicker stems and leaves, and has yellowish-white instead of dark red nectariferous scales (Fröderstöm, 1936; Clausen, 1959). In Jalisco it is known from the Cerro Viejo volcano, the Sierra de Quila, and the Sierra de Tapalpa (1900 -2500 m). It produces white flowers from June to October. *S. griseum* also occurs in the states of Guanajuato, Michoacán, México, and Querétaro.

7. *S. guadalajaranum* S. Watson. This species was discovered by E. Palmer at Rio Blanco (Jalisco) in 1886, and was described by S. Watson in 1887 (see Clausen, 1978). Clausen (l.c.) described *S. guadalajaranum* ssp. *viridifolium* from El Garruño (Jalisco), which, in addition, has also been reported for the state of Zacetacas. In Jalisco *S. guadalajaranum* ssp. *guadalajaranum* is also known from the Cerro Viejo volcano (Cházaro & Machuca, 1992), the Sierra of Tapalpa, and the Cerro La Campana, near Talpa. *S. guadalajaranum* is in bloom from June to August. Most probably it also occurs in the state of Aguascalientes.

8. *S. hintonii* Clausen. Clausen (1981) reported *S. hintonii* for Western Jalisco, but did not give exact localities. A putative report by Sanchez Mejorada (1975) is from Chamela, at the coast of Jalisco. *S. hintonii* also occurs in the states of Durango, Michoacán, and Zacatecas.

9. *S. jaliscanum* S. Watson. The species was discovered by E. Palmer in 1886 in the ravine of Guadalajara, and described by S. Watson in 1887 (see Clausen, 1978). *S. jaliscanum* is a small, annual herb, often forming dense carpets on rocks. It is widely distributed in Jalisco, especially in the region between Guadalajara and Puerto Los Mazos, 10 miles South of Autlán. It produces white flowers during June, July, August, and September. *S. jaliscanum* also occurs in the states of Distrito Federal, Guanajuato, Guerrero, México, Michoacán, Morelos, Nayarit, Oaxaca, and Zacatecas.

10. *S. longipes* Rose. A stonecrop strictly confined to rocky places. J.A. Lomeli found it in Jalisco at Tapalpa along the road to Venustiano Carranza for the first time in 1985 (Cházaro & al., 1994). The whitish-red flowers appear from September to January. *S. longipes* also occurs in the states of Chihuahua, Morelos, México, and Michoacán.

11. *S. machucana* Chazaro & Metzger (in sched.). Endemic to a canyon North of Concepcion de Buenos Aires (Jalisco).

12. *S. moranense* H.B.K. A species from Eastern Mexico. It was first reported for Jalisco (as *S. submontanum,* without exact locality) by Britton & Rose (1905). In addition, Uhl (1983) reported a collection from the South-Western edge of the Sierra Tapalpa, 2130 m (Boutin & Kimnach). Recently, M.J. Cházaro also collected it on the Sierra Tapalpa (Cházaro & al., in press). It flowers in July. *S. moranense* occurs in the states of Chihuahua, Tlaxcala, Distrito Federal, Guanajuato, Hidalgo, México, Michoacán, Puebla, Querétaro, San Luis Potosí, Veracruz, and Zacatecas.

13. *S. multiflorum* Clausen. Described by Clausen (1978) from Los Guayabos ravine, East of Ciudad Guzman (Jalisco). A second locality was discovered on the Sierra Tapalpa and reported by Chazaro & al. (1993). On the Sierra Tapalpa *S. multiflorum* grows on humid rocky cliffs in the vicinity of a waterfall. It has white flowers which bloom in December and January.

14. *S. pringlei* S. Watson. This species was collected by J.N. Rose at San Martin de Bolaños, Sierra Madre (Sierra de los Huicholes) W of Bolaños (Jalisco) in 1897. Fröderström (1935) described this specimen as *S. dispersum,* but Clausen (1984) placed *S. dispersum* in the synonymy of *S. pringlei. S. pringlei* also occurs in the states of Aguascalientes, Chihuahua, Durango, and Hidalogo.

15. *S. semiteres* Rose. This species was reported by Clausen (1978) from the mesophytic tropical forest at the gorge of Los Guayabos Creek near Ciudad Guzman, at about 1775 m (Jalisco). *S. semiteres* also occurs in the state of Durango.

16. *S. tortuosum* Hemsl. The only epiphytic Crassulaceae occurring in Jalisco, it grows on branches of old oak trees, on the Cerro Viejo volcano, the Nevado de Colima, the Sierra de Manantlán, the Sierra de Halo, and the Sierra de Talpa-Cuale. In bloom during February and March. *S. tortuosum* also occurs in the states of Oaxaca, Durango, Guerrero, Michoacán, and México.

17. *S. vinicolor* S. Watson. This species was reported by Clausen (1977) from Jalisco, but without exact locality (Mexican Plateau). *S. vinicolor* also occurs in the states of Sonora, Chihuahua, and Durango.

Villadia Rose

The genus was named after the Mexican naturalist Manuel M. Villada. It is confined to the U.S.A (SW Texas), Mexico, Guatemala, and Peru.

1. *V. batesii* (Hemsl.) Baehni & Macbr. In Jalisco this species is only known from the volcano Nevado de Colima. *V. batesii* is very abundant throughout the Trans-Mexican Volcanic Belt and occurs in the states of Distrito Federal, Hidalgo, Jalisco, México, Puebla, and Tlaxcala.

2. *V. painteri* Rose. A small herb endemic to the Guadalajara Ravine (Jalisco). Dis-

covered independently by J.N. Rose and J. Painter and by C.G. Pringle at the same location in September 1903 (Britton & Rose, 1905). It flowers from July to November.

3. *V. platystyla* (Fröderström) Clausen. Collected by Y. Mexia at Cerro La Bufa, near San Sebastian del Oeste (Jalisco) in 1927, and described by Fröderström (1936) as *Sedum platystylum*. Clausen (1959) transferred it to Villadia. *V. platystyla* is endemic to Jalisco and, in addition to the type locality, is only known from Volcan de Tequila. It is in bloom in January and February.

Phytogeographical affinities of the Crassulaceae of Jalisco - Thiede (this volume) distinguishes three phytogeographical regions within the Crassulaceae of Central and North America, which are based on a cluster analysis of the distribution patterns of some 400 species. Of these regions the "Tropical Mexican Region" (Fig. 1) encompasses tropical and subtropical Mexico, parts of the Southern USA, and Guatemala. Within the Tropical Mexican Region Thiede l.c. distinguishes four subregions, which are by and large congruent with the subprovinces of the "Mexican Highlands Province" of Takhtajan's holarctic "Madrean Region" (Takhtajan, 1986). The state of Jalisco is included in Thiede's "Trans-Mexican Volcanic Belt and Mexican Altiplano Subprovince" of which the Western part is almost congruent with Rzedowski & McVaugh's phytogeographical region "Nueva Galicia" (Rzedowski & McVaugh, 1966).

In figure 1 the total distribution of the 33 species of Crassulaceae occurring in Jalisco is shown on a square degree grid. Most Crassulaceae occurring in Jalisco are restricted to the "Trans-Mexican Volcanic Belt and Mexican Altiplano Subprovince". Six species extend to the North and the South into the "Sierra Madre Occidental Subprovince" and the "Sierra Madre del Sur Subprovince", respectively, and only two to the North-East into the "Sierra Madre Oriental Subprovince". Floristic relations to regions South of the Trans-Mexican Volcanic Belt are sparse and restricted to the Western chain of the Sierra Madre del Sur and the Tehuacán Valley and Oaxaca, notwithstanding the extraordinary species richness of the two latter regions (Thiede, this volume). Affinities to the Northern part of the Sierra Madre Occidental are also sparse, and there are almost no similarities with the Crassulaceae flora of the Sierra Madre Oriental.

Within Jalisco, the highest species richness is evidently found within the "Jalisco Regional Centre" of species richness and endemism as defined by Thiede (this volume), where 73% of all Crassulaceae of Jalisco occur (24 species, including 9 endemic species). This figure clearly emphasizes the importance of the state of Jalisco for the distribution and evolution of the Crassulaceae of Western Mexico. Floristically the Crassulaceae of the "Jalisco Regional Centre" have strong affinities with the species of the "Mexico/ Morelos Centre", the "Mazatlán-Durango highway Centre", the "San Luis Potosí Centre", and the "Hidalgo-Querétaro Centre", which are all situated within the "Trans-Mexican Volcanic Belt and Mexican Altiplano Subprovince" (Thiede, this volume). The Crassulaceae of the "Jalisco Regional Centre" have almost no affinities to floristic centres outside this subprovince.

Comparison of Crassulaceae floras among Mexican states shows that the Crassulaceae flora of Jalisco is most similar to the floras of the neighbouring states, viz. with

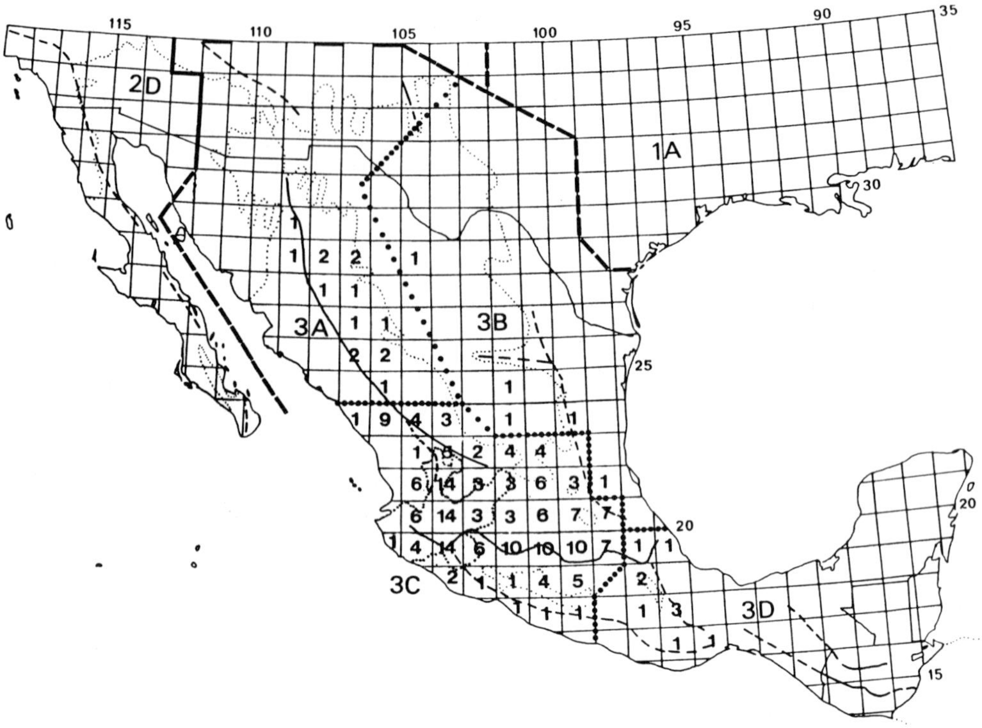

Figure 1. Overall distribution of the 33 Crassulaceae-species occuring in the state of Jalisco (dotted line represents state line), indicated as species number per square degree grid. Phytogeographic delimitation of the northern American Crassulaceae-flora (Thiede, this volume): 1A North American Atlantic Region within the Eastern Temperate Region, 2D Baja Californian Subprovince within the Western Temperate Region, 3 Tropical Mexican Region with 3A Sierra Madre Occidental Subprovince, 3B Sierra Madre Oriental Subprovince, 3C Trans-Mexican Volcanic Belt-Mexican Altiplano Subprovince and 3D Sierra Madre del Sur Subprovince-mountains of the Caribbean Region. Further explanations in the text.

Durango to the North (12 species), Michoacán to the South (12 species), and Guanajuato to the East (8 species). However, it also has much affinity with the floras of species rich states further Eastwards, e.g. with México (11 species) and Hidalgo (9 species). Floristic similarities to more distant states are generally sparse (e.g., Sonora, Veracruz, Oaxaca), and far-reaching phytogeographic affinities are mostly due to a few widely distributed species such as *Echeveria mucronata*, *Sedum jaliscanum*, *S. ebracteatum* and *S. moranense*, which occur in eleven to fourteen of the twenty-two Mexican states.

On the level of the distribution patterns of single species roughly five different phytogeographic elements (chorotypes) can be distinguished:

1) The endemic element includes ten species, one subspecies, and one variety, i.e. *Echeveria chapalensis*, *E. colorata*, *E. elegans* var. *tuxpanensis*, *E. lozanii*, *E. pringlei*, *Graptopetalum fruticosum*, *Sedum grandipetalum*, *S. greggii* ssp. *angustifolium*, *S.*

machucana, *S. multiflorum*, *Villadia platystyla*, and *V. painteri*, and is restricted to the "Jalisco Regional Centre".
2) An element with Northern affinities, which is mainly distributed along the (Southern) Sierra Madre Occidental and reaches its most Southern limit in the state of Jalisco. It includes four species, i.e. *Echeveria dactylifera*, *Sedum guadalajaranum*, *S. semiteres*, and *S. vinicolor.*
3) An element with Eastern affinities, which is represented by *Echeveria fulgens*, *Sedum bourgaei*, and *S. longipes*. It is mainly distributed along the Trans-Mexican Volcanic Belt, and reaches its most Western limit in the state of Jalisco.
4) Another element with Eastern affinities, which is mainly distributed in the Southern part of the Central Mexican Plateau and also reaches its most Western limit in Jalisco. It consists of three species, i.e. *Echeveria agavoides*, *E. elegans*, and *Graptopetalum pachyphyllum.*
5) An element with Southern affinities (Sierra Madre del Sur), which reaches its most Northern limit in Jalisco. It includes *Graptopetalum pentandrum* and *Sedum tortuosum*, though the latter is also known from one locality in the Sierra Madre Occidental.
6) The remaining nine species, i.e. *Graptopetalum amethystinum*, *Sedum fuscum*, *S. griseum*, *S. greggii*, *S. hintonii*, *S. jaliscanum*, *S. moranense*, *S. pringlei*, and *Villadia batesii*, are either widely distributed and occur in all four subprovinces, or are restricted to different combinations of two or three of the chrorotypes.

In summary, these phytogeographic data clearly indicate that the larger part of the Crassulaceae flora of the state of Jalisco is a combination of elements from the Crassulaceae floras of the Southern Sierra Madre Occidental, the Trans-Mexican Volcanic Belt and Southern part of the Central-Mexican Plateau, as well as the Sierra Madre del Sur. In addition it comprises a strong endemic element (c. 30%), which fully justifies future, more extensive studies of this rich flora.

Acknowledgements

I am very grateful to Prof. Charles H. Uhl (Ithaca) who provided me with all his articles as well as valuable criticism on my manuscripts during the past few years. For the same reasons I would like to thank Myron Kimnach (San Marino), Dr. Reid Moran (San Francisco), Dr. Urs Eggli (Zürich), and Dr. Jorge Meyran (México). For companionship and help during the numerous field trips thanks are due to Agustin Flores Macias, Martin Huerta Martinez, Raul Acevedo Rosas, Martin Negrete Aguayo, Jean Metzger, Jose A. Lomeli Sension, Burl Mostul, Patricia Hernández de Cházaro, Arturo García Guerrero, and Jose Antonio Machuca Nuñez. Financial support by the Cactus and Succulent Society of America research fund allowed me to carry out several field trips. Last but not least I wish to thank Elba Lomeli Mijes, Rosa Maria Patiño Beltran, Maria Dolores Alvarez Contreras, and Patricia Moreno Manzo for their help in preparing the manuscript.

References

Britton, N.L. & Rose, J.N. 1905. North American flora, 2, 1. New York Botanical Garden, New York.

Cházaro Basañez, M.J. 1990a. Crassulaceae del centro de Veracruz y zona limitrofe de Puebla, 1. *Cact. Succ. Mex.* **35**: 63-67.

Cházaro Basañez, M.J. 1990b. Crassulaceae del centro de Veracruz y zona limitrofe de Puebla, 2. *Cact. Succ. Mex.* **35**: 83-90.

Cházaro Basañez, M.J. & Flores M., A. 1992. *Graptopetalum pentandrum* ssp. *superbum* Kimnach (Crassulaceae) from the wild. *Cact. Succ. J. (US)* **64**: 187-189.

Cházaro Basañez, M.J., Huerta M., Machuca M., J.A., F.M., Flores M., A., & Hernandez de Chazaro, P. 1992. Notas sobre *Echeveria colorata* E. Walther (Crassulaceae). *Cact. Succ. Mex.* **37**: 35-40.

Cházaro Basañez, M.J., Huerta M., F.M., Patiño B., R.M. & Hernandez de Chazaro, P. 1993. *Sedum multiflorum* R.T. Clausen (Crassulaceae) a poorly known stonecrop. *Cact. Succ. J (US)* **65**: 152-154.

Cházaro Basañez, M.J., Huerta, H., Metzger, J., Vazquez D., M.L. & Hernandez de Chazaro, P. 1994. *Sedum longipes* Rose (Crassulaceae) en Jalisco, Mexico. *Cact. Succ. Mex.* **39**: 18-21.

Cházaro Basañez, M. & Machuca N., J.A. 1992. Notas sobre las cactaceas y otras suculentas del Cerro Viejo y areas circunvecinas, Jalisco (Mexico). *Cact. Succ. Mex.* **37**: 64-70.

Cházaro Basañez, M.J. & Viveros, C. 1991. Nota sobre *Sedum nussbaumerianum* Bitter (Crassulaceae). *Cact. Succ. Mex.* **36**: 43.

Clausen, R.T. 1943. The section *Sedastrum* of *Sedum. Bull. Torrey Bot. Club* **70**: 289-296.

Clausen, R.T. 1959. *Sedum* of the Trans-Mexican Volcanic Belt. Cornell University Press, Ithaca.

Clausen, R.T. 1977. Biennal species of *Sedum* of the Sierra Madre Occidental and Mexican Plateau. *Bull. Torrey Bot. Club* **104**: 209-217.

Clausen, R.T. 1978. *Sedum* - Seven Mexican perennial species. *Bull. Torrey Bot. Club* **105**: 214-223.

Clausen, R.T. 1981. Variation of species of *Sedum* of the Mexican Cordilleran Plateau. Published by the author, Ithaca.

Clausen, R.T. 1984. *Sedum* (Crassulaceae) of the Mexican Cordilleran Plateau. *Gentes Herb.* **12**: 8-48.

Flores M., A. & Cházaro Basañez, M.J. 1993. El status de *Echeveria pringlei* Watson y datos sobre su descubridor. *Cact. Succ. Mex.* **38**: 7-10.

Flores M., A. & Cházaro Basañez, M.J. *G. amethystinum* (in press).

Fröderström, H. 1936. The genus *Sedum* L., 4. *Acta Horti Gothob.* **10**(app.): 1-262.

Gold D.B. 1968. Las cactáceas del estado de Jalisco. *Cact. Succ. Mex.* **8**: 39-41.

Huerta M., F.M. 1993. Crasuláceas y otras succulentas de la Barranca del Nogal, Tapalpa, Jalisco. *Cact. Succ. Mex.* **38**: 38-43.

Kimnach, M. 1980. A revision of *Echeveria colorata* Walther. *Cact. Succ. J. (US)* **52**: 55-63.

Kimnach, M. 1987. A new succulent from Mexico: *Graptopetalum pentrandum* subsp. *superbum. Cact. Succ. J. (US)* **59**: 140-143.

Lomelí, J.A. 1988. *Graptopetalum fruticosum* (Crassulaceae) en el sur de Jalisco. *Cact. Succ. Mex.* **33**: 89-91.

Moran, R. 1963. *Graptopetalum amethystinum* (Rose) Walther. *Cact. Succ. Mex.* **8**: 31-37.

Moran, R. 1964. *Graptopetalum pachyphyllum. Cact. Succ. Mex.* **9**: 9-11.

Moran, R. 1971. *Graptopetalum pentandrum*, a new species from Mexico. *Cact. Succ. J. (US)* **63**: 255-258.

Moran, R. & Uhl, C.H. 1968. *Graptopetalum fruticosum*, a new species from Jalisco, Mexico. *Cact. Succ. J. (US)* **40**: 152-156.

Moran, R. & Uhl, C.H. 1989. *Echeveria chapalensis*, una nueva especie del oeste de México. *Cact. Succ. Mex.* **34**: 27-33.

Reppenhagen, W. 1985. Feldnummern-Verzeichnis W. Reppenhagen 1959-1984. Sonderheft Arbeitskreis Mammillarienfreunde e.V., Osnabrück.

Rzedowski, J. & McVaugh, R. 1966. La vegetación de Nueva Galicia. *Contr. Univ. Mich. Herb.* **9**: 1-123.

Sanchez Mejorada, H. 1975. Un interesante *Sedum* de Chamela, Jalisco. *Cact. Succ. Mex.* **20**: 84-89.

Takhtajan, A. 1986. Floristic regions of the world. University of California Press, Berkeley.

Uhl, C.H. 1983. Chromosomes of Mexican *Sedum*, 4. Heteroploidy in *Sedum moranense*. *Rhodora* **85**: 243-252.

Uhl, C.H. 1989. The hybrid origin of *Echeveria sayulensis*. *Cact. Succ. J. (US)* **61**: 279-284.

Walther, E. 1958. *Echeveria* DC., localidades conocidas en México por estados. *Cact. Succ. Mex.* **3**: 51-54.

Walther, E. 1972. *Echeveria*. California Academy of Sciences, San Francisco.

Table 2. Examples of the contribution of *Crassula* to the Greater Cape Flora: Succulent Karoo Region exclusively

C. alstonii Marl.	*C. nudicaulis* L.
C. ammophila Tölken	var. *herrei* (Friedr.) Tölken
C. atropurpurea (Haw.) Dietr.	*C. numaisensis* Friedr.
var. *cultriformis* (Friedr.) Tölken	*C. pallens* Schönl. & Bak.f.
var. *watermeyeri* (Compton) Tölken	*C. plegmatoides* Friedr.
C. barkleyi N.E. Brown	*C. pseudohemisphaerica* Friedr.
C. brevifolia Harv.	*C. rudolfii* Schönl. & Bak.f.
ssp. *brevifolia*	*C. rupestris* Thunb.
ssp. *psammophila* Tölken	ssp. *commutata* (Friedr.) Tölken
C. columella Marl. & Schönl.	*C. sericea* Schönl.
C. columnaris Thunb.	var. *hottentotta* (Marl. & Schönl.) Tölken
C. cotyledonis Thunb.	*C. sladenii* Schönl.
C. deceptor Schönl. & Bak.f.	*C. subacaulis* Schönl. & Bak.f.
C. elegans Schönl. & Bak.f.	ssp. *erosula* (N.E. Brown) Tölken
C. expansa Dyrand.	*C. subaphylla* (Eckl. & Zeyh.) Harv.
ssp. *pyrifolia* (Compton) Tölken	var. *virgata* (Harv.) Tölken
C. fusca Herre	*C. susannae* Rauh & Friedr.
C. grisea Schönl.	*C. tenuipedicellata* Schönl. & Bak.f.
C. hemisphaerica Thunb.	*C. tetragona* L.
C. hirtipes Harv.	ssp. *lignescens* Tölken
C. macowaniana Schönl. & Bak.f.	*C. tomentosa* Thunb.
C. multiceps Harv.	var. *glabrifolia* (Harv.) Schönl.
C. muscosa L.	*C. umbella* Jacq.
var. *obtusifolia* (Harv.) Tölken	*C. whiteheadii* Harv.
C. namaquensis Schönl. & Bak.f.	*C.* sect. *Arta* (Schönl) Tölken
	C. sect. *Columnares* DC.

Table 3. Species of *Crassula* with a distribution mainly in the Little Karoo (Greater Cape Flora, probably Succulent Karoo Region).

C. arborescens (Mill.) Willd.	*C. pubescens* Thunb.
C. barbata Thunb.	*C. pyramidalis* Thunb.
C. congesta N.E. Brown	*C. rogersii* Schönl.
C. lactea Soland.	*C. rubricaulis* Eckl. & Zeyh.
C. mollis Thunb.	*C. tecta* Thunb.
C. montana Thunb.	

azonal habitats, especially on rocky or shallow ground in mountainous landscapes at higher altitudes. However, some taxa, well adapted to zonal habitats, exist here as well, e.g. *C. crenulata* (Eastern Cape to Eastern Natal) which is a "perennial with finger-like underground tubers producing each year one, rarely two or three erect branches, [] growing on grassy slopes, often under trees" (Tölken, 1985). Due to these problems, it is not possible, to describe the distribution by using names of defined phytochoria, but, instead, single patterns have to be described. One very obvious centre of distribution lies in the Eastern Cape, i.e. the region around Port Elizabeth. At least 9 taxa oc-

Table 11. Contribution of taxa of *Crassula* to the most important phytogeographical patterns.

Phytogeographical Unit	Number of Taxa
Cape Floristic Region exclusively	21
Succulent Karoo Region exclusively	38
Little Karoo	11
Cape Floristic Region & Succulent Karoo Region	16
Nama Karoo Region	11
Wider area along the Indian Ocean	19
Eastern Plateau	10
Tropical North-East	2
Wide distribution area	11

Centres of endemism - If we understand phytochoria as phytogeographical units which are defined by the common occurrence of a number of taxa having roughly the same area of distribution or if, at least, large portions of the borders of the area of distribution are identical, then phytochoria are centres of endemism by definition. Looking at endemism on this level, the figures presented in Table 11 indicate the contribution of *Crassula* to the main phytochoria. As we go below the level of phytochoria, small scale centres of endemism can be identified, not necessarily possessing overlapping or identical areas of distribution, but forming smaller regions, in which above average numbers of local endemics occur in more or less close vicinity, e.g. on single peaks of a mountain chain, rocky outcrops in a plain etc. With respect to *Crassula*, examples for such centres of endemism, which sometimes simultaneously also can form phytochoria of low hierarchical level, are found mainly in the Greater Cape Flora. Here at least *C. alstonii*, *C. brevifolia* ssp. *psammophila*, *C. columella*, *C. fusca*, *C. numaisensis*, *C. rupestris* ssp. *commutata*, and *C. sladenii* can be mentioned as endemics for the Richtersveld, while *C. susannae* is a local endemic of Namaqualand, and *C. alcicornis*, *C. grammanthoides*, and *C. elsiae* are endemics for the Cape Floristic Region. Further Eastwards only very few examples of local endemics are found, e.g. *C. streyi*. Interestingly, the Knersvlakte, generally an important centre of endemism, is not of importance for *Crassula*, perhaps because of the high soil salinity (see chapter on ecology).

In comparison to the regions further East the Greater Cape Flora has in general more species with a small size distribution area.

Phytogeographical patterns of the sections of *Crassula* - In order to produce a more objective basis for the discussion of the phytogeographical patterns of *Crassula*, the results of statistical analysis shall be presented. Data exploration by multivariate statistics has become a well-known tool in vegetation analysis. The grid of geographical degree squares and their *Crassula* species composition can be analysed in parallel to vegetation relevés in order to detect the structure of similarities and dissimilarities in the database. If the species compositions of *Crassula* in the degree squares are used as variables in cluster analysis (SPSS, Wards method) the following result is produced rock habitats in a temperate climate. Both habitat preferences are still valid for the majority of species. Even in those species which have adapted to fine-textured soils under desert conditions, the heritage of the primary habitat preference is still existing, as demonstrated by their preference for soils of low salinity and pH.

Furthermore, only very few species of *Crassula*, e.g. *C. rupestris*, have evolved the ability to become dominant in zonal habitats. Most species build up small populations only and the individual populations may occur far apart. This pattern is very characteristic for *Crassula* as a comparison with Mesembryanthemaceae will make evident. Members of the latter family can be of almost identical growth form, but are usually able to occur in large numbers of individuals in zonal habitats, i.e. on deep fine-textured soils with a high salinity and high pH. Both adaptive syndromes can result in similar numbers of species per square kilometer (Jürgens, 1986), while the number of individuals of Mesembryanthemaceae can be 1000 times larger than the number of individuals of *Crassula*.

In this context it is remarkable, that those species of *Crassula* which are able to produce relatively dense populations with many individuals, e.g. *C. ovata*, *C. argentea*, *C. brevifolia*, and *C. rupestris*, are plants which have evolved relatively large growth-forms, i.e. large chamaephytes or nanophanerophytes. This fact makes one aware that small to very small succulent growth forms are well adapted to habitats which are characterized by regular availability of very small quantities of water. However, this growth-form allows survival only in habitats with a low competition pressure.

The phytogeographical distribution of *Crassula* within Southern Africa follows very common patterns, which in a similar form are also found in Mesembryanthemaceae (Jürgens, 1986, 1991), *Pelargonium*, and *Anacampseros*. One of the similar patterns is a strong differentiation into taxa of inland North-East and North in opposition to taxa of the coastal regions with strong emphasis on the winter rainfall region, which is linked to smaller or larger parts of the adjacent Indian Ocean coastal region. In the North-Western Cape Province and in Southern Namibia the differentiation into winter rainfall taxa and summer rainfall taxa is strongly developed. Analysis via cladistic phytogeography for *Anacampseros* (Gerbaulet, 1992a,b) led to the result that a differentiation into tropical and temperate taxa formed the first step in the evolution of *Anacampseros*. Subsequently the tropical and temperate taxa independently developed an arid-adapted taxon.

Another very frequent pattern is the differentiation into winter rainfall taxa and summer rainfall taxa. At a family level the distribution area of the Mesembryanthemaceae is similar to the area of *Crassula*. But at the next lowest hierarchical levels the taxa are well differentiated into winter rainfall taxa and summer rainfall taxa (Jürgens, 1986; Hartmann, 1991). This observation is in harmony with the comparatively young (4 to 5 million years B.P.) development of the winter rainfall climate (Axelrod & Raven, 1978).

If these models are applied to the construction of a hypothesis of evolution in *Crassula*, similar patterns can be interpreted in a similar way. Then the primary distribution of *Crassula* was confined to edaphically arid rock habitats along the Southern margin, or most temperate part, of the continent. Further evolution probably did not

Berger and *S.* sect. *Seda genuina* series *Propontica* Berger, which are based on vegetative characters. Fröderström (1932) also distinguished two series in the *S. spurium* group, but used the shape of the ripe follicles (orthocarpic versus kyphocarpic) to delimit the series. Borissova (1939) united the series in *S.* sect. *Sedum* subsect. *Spathulata.* Ohba (1978) raised Borissova's section to the rank of subgenus and distinguished to sections based on the shape of the ovaries. Grulich (1984) proposed the new genus *Asterosedum* for this taxon. Grulich (1984) and 't Hart (1984) independently added the annual *Sedum stellatum* L., the lectotype of *Phedimus* to the *S. spurium* group.

't Hart et al. (1993) revealed that the species of sect. *Stolonifera* and sect. *Spathulata* of *Sedum* subgen. *Spathulata* differ significantly in basic chromosome number, amount of DNA per nucleus, and karyotype characteristics. Within *Phedimus* the variation in cytological and morphological characters, e.g. the shape of the follicles, seems sufficiently large to separate *P. spurius* (Bieb.) Grulich (≡ *Sedum spurium* Bieb.) at sectional level from the remaining species (Ohba, this paper).

The new genus *Balfouria* and the problem of disparate groups in *Sedum*

Hamet (1912) described *Sedum balfourii* based on a Forrest collection from the Kari pass, Yangtse-Mekong divide, in Eastern Tibet, and later included *S. mossii* Hamet, *S. orichalcum* W.W. Smith, and *S. banlanense* Limpricht in the synonymy of *S. balfourii* (Hamet 1929). Fröderström (1931) followed Hamet (1929) and classified *S. balfourii* in the *Umbilicoides* group of his section *Asiatica Genuina Orthocarpia* of *Sedum.*

Sedum balfourii is a perennial herb with a conspicuous rosette, 3-7 cm in diameter, consisting of 20-40 leaves. Rosulate leaves narrowly oblong to oblong-lanceolate, glabrous, 1-3 cm long, 0.5-1.2 cm wide, apex obtuse or acute, margin nearly entire to slightly ciliate, base dilatated and widely clasping. Flowering stems arising from the axils of rosulate leaves, 10-30 cm long, glabrous and smooth with alternate leaves; cauline leaves similar to rosette leaves in shape but smaller, 1.5-2.6 cm long, 3.5-7.0 mm wide. Flowers in a terminal cyme with three primary axes and leafy bracts, hermaphrodite, obdiplostemonous, 5- or 6-merous, shortly pedicellate. Sepals basally connate for c. 0.5 mm, lanceolate to triangular-ovate, 2.5-3.5 mm long, 1.0-1.5 mm wide, fleshy, glabrous, acute. Petals free, yellow, ovate to lanceolate, 5.5-6.8 mm long, 2.0-2.5 mm wide, mucronulate, erect at anthesis. Stamens shorter than petals, with subulate filaments and oblong, 0.8-1.3 mm long anthers. Carpels 4-5 mm long, basally connate for c. 0.5 mm, suberect at anthesis, ventrally not gibbous, with a slender style 1.5 mm long and submarginal placentas with 30-50 ovules in each locule.

Praeger (1917, 1921) noted that *Sedum balfourii* has rosulate leaves which are connected to the caudex with a broad and clasping base exactly as in the *S. primuloides* group of *S.* sect. *Rhodiola* (L.) Scop., and he thus established a *Brevicaulis* group comprising *S. durisii* Hamet, *S. hobsonii* Prain, *S. balfourii*, *S. karpelesae* Hamet, *S. levii* Hamet, *S. praegerianum* W.W. Smith, and, tentatively, *S. mossii.* Berger (1930) formally described a series *Brevicaulis* of *S.* sect. *Rhodiola* subsect. *Primuloides* (Praeger)

Berger.

In a previous contribution (Ohba, 1978), I have already indicated the heterogeneous nature of Berger's *Sedum* series *Brevicaulis* which comprises the following taxa (Berger's name given in brackets): *Rhodiola humilis* (Hook. f. & Thom.) Fu (*Sedum levii* and*S. karpelesae*), *Rhodiola hobsonii* (Prain) Fu (*Sedum hobsonii* and *S. praegerianum*), *Rosularia alpestris* (Karelin & Kiriloff) A. Boriss. (*Sedum durisii*) and *S. balfourii*. At the same time I proposed the monotypic *S.* subgenus *Balfouria*, which is characterized by conspicuous rosettes as in *Rosularia* (Ohba, 1978).

Already Hamet (1929) pointed out that *Sedum balfourii* is similar to *Rosularia* in several points. Although the characters of the follicles, seeds and testa are presently not available, it is clear that *S. balfourii* differs from *Rosularia* as the former is completely glabrous and, moreover, has free petals as has been pointed out by Eggli (1988). *S. balfouri* differs greatly from the species of *Rhodiola*. It lacks the scaly rhizomes of the latter and, in contrast, has yellow petals. *Orostachys* also differs greatly from *S. balfourii* by the stipitate carpels and the basally connate petals. *Sinocrassula* differs from *S. balfourii* by its haplostemonous flowers and the position of its inflorescences (terminal to monocarpic rosettes). *Meterostachys* finally differs from *S. balfourii* by its basally connate petals and sepals forming a conspicuous tube.

Except for the species of *Sedum* sect. *Cepaea* W.D.J. Koch, no other Eurasian *Sedum* taxon has a similar combination of conspicuously glabrous rosettes and yellow petals as *S. balfourii*. However, the taxa classified in *S.* sect. *Cepaea*, viz. *S. lampusae* (Kotschy) Boissier, *S. microstachyum* (Kotschy) Boissier, *S. cyprium* A.K. Jackson & Turrill, etc., differ from *S. balfourii* in their monocarpic habit, the terminal thyrsoid-paniculate inflorescences, and the spathulate or linear-spathulate rosulate leaves.

Sedum balfourii does not show any synapomorphic features with other taxa of Sedoideae, and it appears sufficiently distinct from all genera to be treated tentatively as a distinct genus, *Balfouria*.

Future attempts to improve the classification of the Sedoideae must deal adequately with the numerous disparate groups within the large genus *Sedum*, now more than ever before recognized to be heterogeneous. One possible solution is the separation of the various disparate groups as genera in their own right. Another possibility is the acceptance of the genus *Sedum* as a composite or paraphyletic taxon. In my opinion, the genus *Sedum* (excluding *Balfouria*, *Asterosedum* and *Aizopsis*) should be revised to include only clade 7 or clades 5 to 7 of the phylogeny of the Crassulaceae by van Ham & 't Hart (1994).

Synopsis of supraspecific taxa of the Asian Sedoideae

In 1978 I proposed a classification of the Sedoideae which was mainly applicable to the Asian species of the subfamily (Ohba, 1978). I accepted 35 infrafamilial (generic and infrageneric) taxa. The boundaries of each taxon were clear, and there were no species holding an intermediate position. Though the treatment of *Sedum* still seems inadequate, the delimitation of the genus is beyond my basically regional approach to its taxonomy.

A revised tentative classification of the Asian Sedoideae which closely follows my previous system is presented below. Disparate species or species groups are classified as genera or infrageneric taxa, i.e. *Sinocrassula*, *Meterostachys*, *Balfouria*, *Hylotelephium* sect. *Populisedum*, *Orostachys* sect. *Schoenlandia*, etc. With very few exceptions, these taxa are mono- or oligotypic, and their geographical distribution ranges from the Himalaya to South-Western China through Southern Tibet and Northern Burma. At present, these regions have very rich floras which are maintained by a mild climate with plentiful rainfall. It is interesting that the disparate taxa of Asian Sedoideae are by and large confined to these humid climatic regions.

Despite their morphological distinctness, the phylogenetic relationship and systematic position of *Sinocrassula*, *Meterostachys*, and *Pseudosedum* are still unresolved.

Conspectus of Asian Sedoideae

Hylotelephium H. Ohba in Bot. Mag. Tokyo 90, 46. 1977. Type, *Hylotelephium telephium* (L.) H. Ohba (≡ *Sedum telephium* L.).

sect. *Hylotelephium*

sect. *Sieboldiana* (H. Ohba) H. Ohba, stat. nov.

Basionym: *Hylotelephium* series *Sieboldiana* H. Ohba in J. Fac. Sci. Univ. Tokyo, Sect. 3, 12, 165. 1978. Type, *Hylotelephium sieboldii* (Hook.) H. Ohba (≡ *Sedum sieboldii* Hook.).

sect. *Populisedum* (A. Berger) H. Ohba in J. Fac. Sci. Univ. Tokyo, Sect. 3, 12, 165. 1978. Type, *Hylotelephium populifolium* (Pall.) H. Ohba (≡ *Sedum populifolium* Pall.).

Orostachys (DC.) Fisch. in Cat. Jard. Gorenki 99. 1808. Type, *Orostachys malacophyllus* (Pall.) Fisch. (≡ *Cotyledon malacophylla* Pall.).

sect. *Orostachys*

subsect. *Orostachys*

subsect. *Appendiculatae* (Boriss.) H. Ohba in J. Fac. Sci. Univ. Tokyo, Sect. 3, 12, 160. 1978. Type, *Orostachys spinosus* (L.) C.A. Meyer (≡ *Cotyledon spinosa* L.).

sect. *Schoenlandia* H. Ohba in J. Fac. Sci. Univ. Tokyo, Sect. 3, 12, 160. 1978. Type: *Orostachys schoenlandii* (Hamet) H. Ohba (≡ *Sedum schoenlandii* Hamet.).

Aizopsis Grulich in Preslia 56, 35. 1984. Type, *Aizopsis aizoon* (L.) Grulich (≡ *Sedum aizoon* L.).

Phedimus Rafinesque in Amer. Monthly Mag. Crit. Rev. 1, 438. 1817. Lectotype, *Phedimus stellatus* (L.) Rafin. (≡ *Sedum stellatum* L.) ('t Hart in Fl. Medit. 1, 52. 1991).

sect. *Phedimus*

sect. *Spathulata* (Boriss.) H. Ohba, comb. nov.
Basionym: *Sedum* subsect. *Spathulata* Boriss. in Novit. Syst. Pl. Vasc. Acad. Sci. USSR 6, 117. 1969. Type, *Phedimus spurius* (Bieb.) 't Hart (≡ *Sedum spurium* Bieb.).

Rhodiola L. in Sp. Pl. 1035. 1753. Type, *Rhodiola rosea* L.
subgen. *Primuloides* (Praeger) H. Ohba in J. Fac. Sci. Univ. Tokyo, Sect. 3, 12, 183. 1978. Type, *Rhodiola primuloides* (Franch.) Fu (≡ *Sedum primuloides* Franch.).
sect. *Hobsonia* (H. Ohba) H. Ohba in J. Fac. Sci. Univ. Tokyo, Sect. 3, 12, 184. 1978. Type, *Rhodiola hobsonii* (Hamet) Fu (≡ *Sedum hobsonii* Hamet).
sect. *Primuloides*
sect. *Smithia* (H. Ohba) H. Ohba in J. Fac. Sci. Univ. Tokyo, Sect. 3, 12, 185. 1978. Type, *Rhodiola smithii* (Hamet) Fu (≡ *Sedum smithii* Hamet).
subgen. *Crassipedes* (Praeger) H. Ohba in J. Fac. Sci. Univ. Tokyo, Sect. 3, 12, 185. 1978. Lectotype, *Rhodiola wallichiana* (Hook.) Fu (≡ *Sedum crassipedes* Hook. f. & Thoms.) (H. Ohba, 1978).
subgen. *Clementsia* (Rose) H. Ohba in J. Fac. Sci. Univ. Tokyo, Sect. 3, 12, 186. 1978. Type, *Rhodiola rhodantham* (A. Gray) Jacobsen (≡ *Sedum rhodanthum* A. Gray).
subgen. *Rhodiola*
sect. *Pseudorhodiola* (Diels) H. Ohba in J. Fac. Sci. Univ. Tokyo, Sect. 3, 12, 187. 1978. Lectotype, *Rhodiola yunnanensis* (Franch.) Fu (≡ *Sedum valerianoides* Diels) (H. Ohba, 1978).
sect. *Rhodiola*
sect. *Prainia* (H. Ohba) H. Ohba in J. Fac. Sci. Univ. Tokyo, Sect. 3, 12, 189. 1978. Type, *Rhodiola prainii* (Hamet) H. Ohba (≡ *Sedum prainii* Hamet).
sect. *Chamaerhodiola* (Fisch. & C.A. Mey.) Boriss. in Komarov, Fl. USSR 9, 39. 1939. Type, *Rhodiola quadrifidam* (Pall.) Fisch. & C.A. Meyer (≡ *Sedum quadrifidum* Pall.).

Sedum L., Sp. Pl.: 430 (1753). Lectotype: *Sedum acre* L. (Hitchcock in Prop. Brit. Bot. 156. 1929).

Prometheum (A. Berger) H. Ohba in J. Fac. Sci. Univ. Tokyo, Sect. 3, 12, 168. 1978. Type, *Prometheum sempervivoides* (Bieb.) H. Ohba (≡ *Sedum sempervivoides* Bieb.).

Rosularia (DC.) Stapf in Curtis's Bot. Mag. 149, t. 8985, 1923. (in adnota). Lectotype, *Rosularia sempervivum* (Bieb.) Stapf (≡ *Cotyledon sempervivum* Bieb.) (Borissova in Nov. Syst. Pl. Vasc. Acad. Sci. USSR 6, 119. 1969).

Balfouria (H. Ohba) H. Ohba, stat. nov.
Basionym: *Sedum* subgen. *Balfouria* H. Ohba in J. Fac. Sci. Univ. Tokyo, Sect. 3, 12, 177. 1978. Type, *Balfouria balfourii* (Hamet) H. Ohba comb. nov. (≡ *Sedum balfourii*

Hamet in Notes Roy. Bot. Gard Edinburgh 5, 116. 1912).

Sinocrassula A. Berger in Engler & Prantl, Nat. Pflanzenfam. ed. 2, 18a, 463. 1930. Lectotype, *Sinocrassula indica* (Decne.) A. Berger (≡ *Crassula indica* Decne.) (H. Ohba in J. Fac. Sci. Univ. Tokyo, Sect. 3, 12, 170. 1978).

Meterostachys Nakai in Bot. Mag. Tokyo 49, 74. 1935 (in adnot.). Type, *Meterostachys sikokiana* (Makino) Nakai (≡ *Sedum sikokianum* Makino).

Pseudosedum (Boiss.) A. Berger in Engler & Prantl, Nat. Pflanzenfam. ed. 2, 18a, 465. 1930. Type, *Pseudosedum lievenii* (Ledeb.) A. Berger (≡ *Umbilicus lievenii* Ledeb.)

References

Berger, A. 1930. Crassulaceae. In Engler, A. & Prantl, K. (eds.), Nat. Pflanzenfam., ed. 2, **18a**, 352-483. W. Engelmann, Leipzig.

Borissova, A.G. 1939. Crassulaceae. In Komarov, V. L. (ed.), Flora of USSR **9**, 8-134, 471-186. Izdatel'stvo Akademii Nauk SSSR, Leningrad.

Eggli, U. 1988. A monographic study of the genus *Rosularia* (Crassulaceae). *Bradleya* **6**(Suppl.): 1-119.

Fröderström, H. 1931. The genus *Sedum* L. A systematic essay. 2. *Acta Hort. Gothob.* **6**(app.): 1-111.

Fröderström, H. 1932. The genus *Sedum* L. A systematic essay. 3. *Acta Hort. Gothob.* **7**(app.): 1-125.

Fu, S. H. & Fu, K. T. 1984. Crassulaceae. In Flora Reipublicae Popularis Sinicae **34**, 31-220. Science Press, Beijing. (in Chinese)

Grulich, V. 1984. Generic division of Sedoideae in Europe and adjacent regions. *Preslia* **56**: 29-54.

Ham, R.C.H.J. van & Hart, H. 't. 1994. Phylogenetic relationships in the Crassulaceae inferred from chloroplast DNA restriction-site variation. In Ham, R.C.H.J. van, Phylogenetic Implications of Chloroplast DNA Variationin the Crassulaceae, 83-105. Ph.D. Thesis, Utrecht.

Hamet, R. 1912. Plantae Chinenses Forrestianae. Enumeration and description of species of *Sedum*. *Notes Roy. Bot. Gard. Edinburgh* **5**: 115-122.

Hamet, R. 1929. Contribution à l'étude phytogéographique du genre *Sedum*. *Candollea* **4**: 1-52.

Hart, H. 't. 1984. The white-flowered European *Sedum* species. 3. The systematic position of *S. stellatum* L. *Proc. Kon. Ned. Akad. Wet., Ser. C,* **87**: 401-411.

Hart, H. 't Tomlik, A. & Alpinar, K. 1993. Biosystematic studies in *Sedum* (Crassulaceae) from Turkey. 4. The cytology of *Sedum* subsect. *Spathulata* Boriss. *Acta Bot. Neerl.* **42**: 289-298.

Ohba, H. 1978. Generic and infrageneric classification of the Old World Sedoideae (Crassulaceae). *J. Fac. Sci. Univ. Tokyo, Sect. 3,* **12**: 139-198.

Ohba, H. 1980. A revision of the Asiatic species of Sedoideae (Crassulaceae). 1. *J. Fac. Sci. Univ. Tokyo, Sect. 3,* **12**: 337-405.

Ohba, H. 1981. A revision of the Asiatic species of Sedoideae (Crassulaceae). 2. *J. Fac. Sci. Univ. Tokyo, Sect. 3,* **13**: 65-119.

Ohba, H. 1982. A revision of the Asiatic species of Sedoideae (Crassulaceae). 3. *J. Fac. Sci. Univ. Tokyo, Sect. 3,* **13**: 121-174.

Ohwi, J. 1953. Flora of Japan. Shibundo, Tokyo. (in Japanese)

Praeger, R.L. 1917. On the affinities of *Sedum Praegerianum* W.W. Sm., with a tentative classification of the section *Rhodiola. Trans. Bot. Soc. Edinburgh* **27**: 107-119.

Praeger, R.L. 1921. Some Asiatic *Sedums* in the Edinburgh herbarium. *Notes Roy. Bot. Gard. Edinburgh* **13**: 67-101.

Wassmer, A. 1955. Vergleichend-morphologische Untersuchungen an den Blüten der Crassulaceen. Diss. Univ. Zürich (*Arb. Inst. Allg. Bot. Univ. Zürich, Ser. A,* **7**).

C H A P T E R 1 0

Infrafamilial and generic classification of the Crassulaceae

Henk 't Hart

Abstract

A new infrafamilial classification of the Crassulaceae is presented which is primarily based on chloroplast DNA characters. The family (1,300 - 1,400 species in about 33 genera) is divided into two subfamilies, the Crassuloideae with c. 250 species in two genera, and the Sedoideae with about 1,100 species. The Sedoideae are further divided into two tribes, the Kalanchoeae with c. 250 species in 5 genera and the Sedeae. The latter tribe is divided into two subtribes, the Telephiinae with c. 150 species in 8 genera and the Sedinae with c. 700 species in 18 genera. About half the number of species of the Sedinae belong to the large genus *Sedum* sensu stricto (c. 350 species). The other genera of this subtribe vary in size from very large (*Echeveria* with 150-200 species), to medium (50-40 species, *Aeonium*, *Dudleya*, and *Sempervivum*), and small to monotypic (less than 15 species). CpDNA phylogenies unequivocally indicate *Sedum* to be paraphyletic. The other genera of the Sedinae all seem to have evolved from *Sedum* like ancestors. The cpDNA phylogenies indicate that the Sedinae comprises four more or less distinct lineages each of which contains representatives of *Sedum*. The *Sedum* species of the molecularly very distinct *Acre*-clade, which is also morphologically (ornamentation testa) and phytochemically (alkaloids) distinct, are classified in *S.* subgen. *Sedum*, whereas the rest of the genus is placed in *S.* subgen. *Gormania*.

Introduction

The family Crassulaceae comprises 1,300 - 1,400 species of leaf succulents which predominantly occur in (semi-)arid and rocky (mountainous) habitats of the temperate and subtropical zones of the Northern Hemisphere and Southern Africa. Whereas the family is generally considered to be a very "natural" group, the infrafamilial classification and the delimitation of a majority of Crassulaceae genera are still controversial and subject to an ongoing debate ('t Hart & Eggli, this volume).

The most authoritative and widely accepted infrafamilial classification of the Crassulaceae was written by Berger (1930) for the second edition of Engler & Prantl's series of monographs "Die natürlichen Pflanzenfamilien". In this account Berger recognised 6 subfamilies and 33 genera. The genera vary in number of species along a Willis's hollow curve of distribution (Willis, 1922), with one very large genus (*Sedum*, 300 - 500 species), three large genera (100 - 250 species), several genera with 20 to 50 species, and a score of smaller and monotypic genera. The subfamilies are conceived to contain one of the large genera each, i.e. *Crassula*, *Kalanchoe*, *Cotyledon*, *Sempervivum*, *Sedum*, and *Echeveria*. Berger's (1930) taxonomic treatment of the Crassulaceae is primarily a traditional, straightforward hierarchical classification which owing to its comprehensiveness has proved to be of great practical value, though it is generally acknowledged that substantial parts are highly artificial and in need of revision (Uhl, 1961, 1963; Clausen, 1959, 1975; Ohba, 1978; 't Hart, 1982, 1991; Grulich, 1984; Eggli, 1988).

In the last six decades relationships within the Crassulaceae have been studied at all taxonomic levels with a wide variety of methods (Mauritzon, 1933; Quimby, 1939; Uhl, 1948, 1992; 1955; Wassmer, 1955; Jensen, 1966, 1968; Hideux, 1981; 't Hart & Koek-Noorman, 1989; 't Hart, 1991; Pilon-Smits, 1992; Knapp, 1994; van Ham, 1994, this volume). These studies have not resulted in a new or improved infrafamilial classification so far, except for Thorne's (1983) proposal to reduce the number subfamilies to three. Of the new approaches mentioned only Mauritzon's (1933) embryological study, Hideux's (1981) pollen morphological study, van Ham's (1994) molecular analysis, and Knapp's (1994) inventory of seed ornamentation types produced comprehensive surveys of relationships at the family level. These 4 studies agree in the conclusion that the major division in the Crassulaceae is between the haplostemonous and obdiplostemonous taxa or between Berger's (1930) Crassuloideae and the rest of the family. As the results of the molecular studies (van Ham, 1994, this volume) comprise by far the largest, most detailed, and reliable data sets so far available for this family, they can be used as reference to adapt the infrafamilial and generic classification of the Crassulaceae according to the prerequisites determined by these studies.

Chloroplast DNA phylogeny of the Crassulaceae

The phylogenetic trees of the Crassulaceae based on chloroplast DNA restriction-site analysis presented by van Ham (1994, this volume) comprise 44 species of 19 genera representing all six of Berger's subfamilies, but for a more comprehensive study of intergeneric relationships within the family the taxonomic sampling needs to be extended. In an introductory study of the molecular evolution of noncoding regions of the chloroplast genome van Ham & al. (1994) concluded that sequence analysis of an

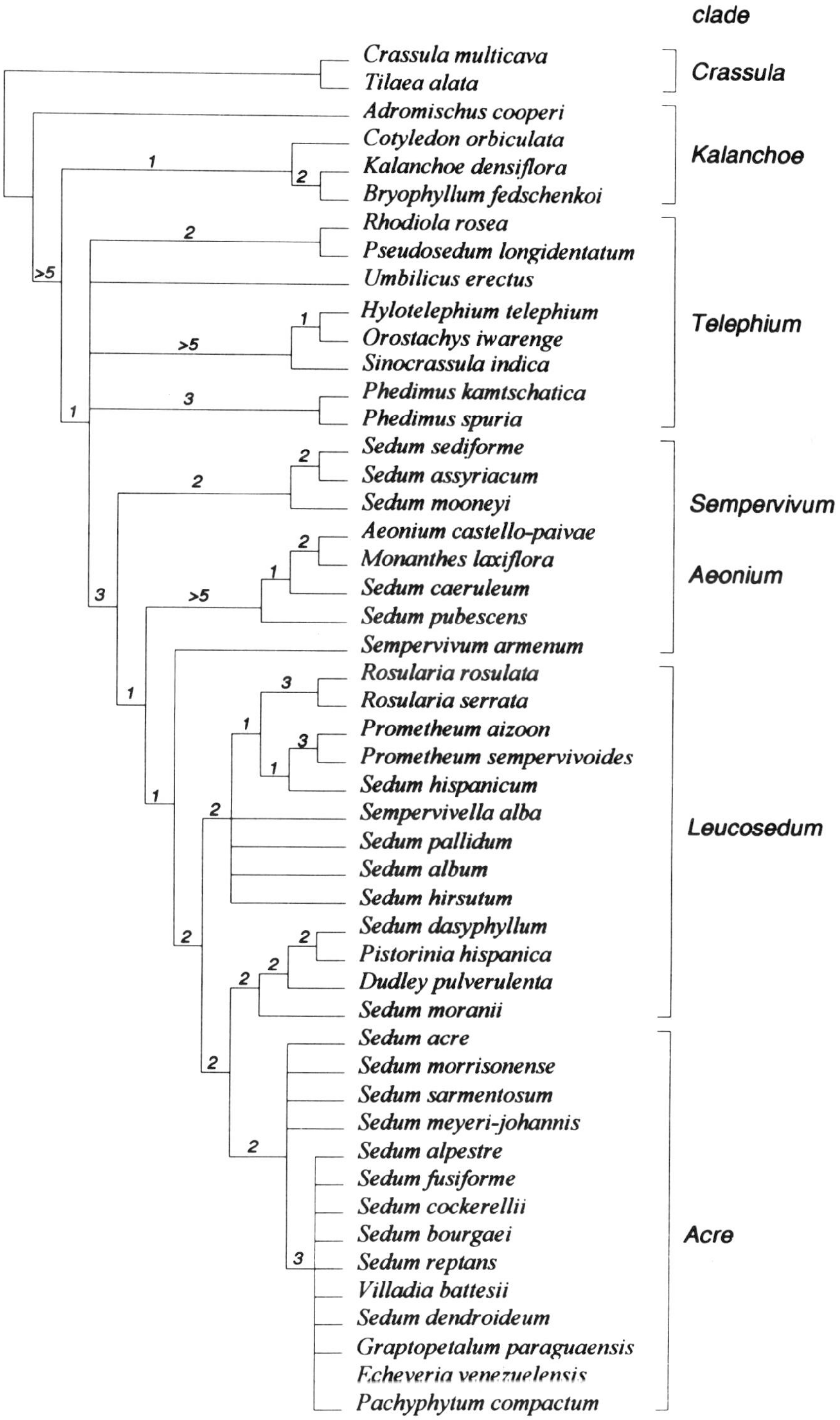

Figure 1. Strict consensus Wagner tree (471 steps, consistency 0.59, informative sites only) based o substitutions of the trnL (UAA) - trnF (GAA) intergenic chloroplast DNA spacer of 49 Crassulacea species of 26 genera, using *Crassula* an *Tillaea* as outgroups. Decay values are indicated above th branches. The clades distinguished by van Ham (1994, this volume) are also indicated.

intergenic spacer between the trnL and trnF genes can also confidently be used to infer phylogenetic relationships within the Crassulaceae. Cladistic analysis of the aligned sequences of the intergenic region between the trnL (UAA) exon and trnF (GAA) gene of 12 Crassulaceae resulted in a phylogeny matching the trees based on restriction-site variation of the same taxa.

First we extended the sample of 12 species used by van Ham & al. (1994) to match the sample of 44 species (19 genera) of van Ham's (1994, this volume) chloroplast DNA restriction-site variation study. Although the portion of the chloroplast genome sampled for sequence analysis is much smaller than sample used in van Ham's restriction-site analysis, the trees found are basically identical to the Wagner tree inferred from restriction-site variation ('t Hart & al., unpublished results). However, these trees are less resolved (containing several polytomies) and robust which is due to the restricted size of the sequences used.

For the present study we extended the sample used by van Ham (1994, this volume) in the study of the family with species of another 6 genera and in the final analysis the size of the sample comprised 49 species of 26 genera. Figure 1 presents the strict consensus tree is of the shortest trees resulting from a heuristic search under Wagner parsimony, using *Crassula* and *Tillaea* as outgroups. With respect to the major lineages this tree is by and large identical to a Neighbor-Joining tree of the same data set. The general topology and major lineages fully agree with van Ham's results. However, some clades are less well-supported or paraphyletic, but these were usually weakly supported in van Ham's (1994) study too. The *Kalanchoe*-clade is paraphyletic with *Adromischus* at its base in a sister group-position to the rest of the family. The *Telephium*-clade is a polytomy with a strongly supported clade comprising *Hylotelephium*, *Orostachys*, and *Sinocrassula* and an also rather well-supported, monophyletic clade with one species of each of the two subgenera of *Phedimus*. The *Sempervivum*-clade and *Aeonium*-clade are combined in a weakly supported, paraphyletic assemblage. The *Sempervivum*-clade is split in a *Sempervivum* branch and a weakly supported clade with two Eurasian and one African species of *Sedum*. The two branches of the former *Sempervivum*-clade are placed on either side of a still strongly supported monophyletic *Aeonium*-clade. The *Leucosedum*-clade and the *Acre*-clade are both paraphyletic and their four branches are only weakly supported. The two branches of the *Leucosedum*-clade comprise taxa from different (sub)continents. The basal branch contains exclusively Eurasian species and comprises two relatively well-supported monophyletic clades, i.e. *Rosularia* and *Prometheum*. The other branch is the sister group of the *Acre*-clade. It comprises the West Mediterranean *Pistorinia* and *Sedum dasyphyllum* L. as well as a *Dudleya* and *Sedum moranii* R.T. Clausen from the Western US. The *Acre*-clade consists of two polytomies of which the terminal one comprises the Eurasian *S. alpestre* Vill., *S. fusiforme* Lowe from Madeira, and 8 American-Mexican taxa, whereas the East Asian (*Sedum morrisonensis* Hayata and *S. sarmentosum* Bunge), African (*S. meyeri-johannis* Engler) and a second Eurasian *Sedum* species (*S. acre*) are part of the basal polytomy of this clade.

Classification of the Crassulaceae

The infrafamilial and generic classification of the Crassulaceae which incorporates the results of the phylogenetic analyses of the chloroplast DNA variation is presented in the Conspectus. A summary of the infrafamilial classification is presented in Table 1. The latter is primarily based on the analysis of restriction-site variation of 44 species of 16 genera (van Ham, 1994, this volume), whereas the arrangement of the genera proposed in the conspectus is primarily based on the sequence analysis of the chloroplast trnL - trnF intergenic spacer (Fig. 1).

Table 1. Origin and numbers of species and genera of the infrafamilial taxa of the Crassulaceae.

Taxon	Species	Genera	Origin
subfam. Crassuloideae	c.250	2	S Africa
subfam. Sedoideae			
tribus Kalanchoeae	c.250	5	S Africa
tribus Sedeae			
subtribus Telephiinae	c.150	8	Asia
subtribus Sedinae	c.700	18	N Hemisphere

The basal split in the Crassulaceae is between the predominantly African Crassuloideae and the rest of the family and separates the haplostemonous taxa with opposite leaves from the taxa which never have this unique combination of characters. The Crassuloideae are identical with the strongly supported, monophyletic *Crassula*-clade of van Ham's (1994, this volume) chloroplast DNA phylogeny. This classification is strongly supported by the embryological (Mauritzon, 1933) and pollen morphological characters (Hideux, 1981). The cytological uniformity of the Crassuloideae (basic numbers x=7 and 8) is further evidence for the monophyly of this subfamily (Merxmüller & al., 1971). The rest of the family is united in subfamily Sedoideae. This taxon is divided into two tribes, the African Kalanchoeae and the Sedeae which is confined to the Northern hemisphere.

Tribus Kalanchoeae is congruent with van Ham's (1994, this volume) *Kalanchoe*-clade and comprises Berger's (1930) subfamily Kalanchoideae and the Southern African members of Berger's subfamily Cotyledonoideae. It is a strongly supported clade in the chloroplast DNA phylogenies and the monophyly of this taxon is further supported by embryological (Mauritzon, 1933) and pollen morphological characters (Hideux, 1981). The unique combination of 4- or 5-merous flowers with distinctly connate petals, costate seeds, and usually flat, crenate or dentate, often petiolate and decussate leaves characterize the tribus Kalanchoeae. Anatomical features of the shoots (Jensen, 1966, 1968) support the postulated affinities between Berger's Kalanchoideae and Southern African Cotyledonoideae. Bufadiënolides (steroïds) have been reported for Madagascan *Bryophyllum* species (Capon & al., 1986; Yamagisha & al., 1989). It seems worthwhile to investigate the taxonomic significance of these cytotoxic compounds (Wagner & al., 1985) as some *Cotyledon* species are also indicated to be very toxic (Berger, 1930).

Tribus Sedeae is again divided into two subtribes, the predominantly Asian Telephiinae and the Sedinae which occur on all three (sub)continents of the Northern hemisphere.

Subtribus Telephiinae comprises the *Telephium*-clade (van Ham, 1994, this volume). Almost all taxa of this subtribus have obdiplostemonous, 5-merous flowers with free petals and have been attributed to *Sedum*, either as subgenus, section, or series (Ohba, 1978). With only few exceptions the species of this subtribus have flat and often crenate or dentate leaves and they have also been referred to as the "flat-leaved, Asian Sedum species" ('t Hart, 1984, 1991). Tubers, tuberous roots, or woody or thickened rhizomes, either monopodial or sympodial, are very characteristic for this group. Seeds with a multipapillate testa have only been reported for species of this subtribe so far ('t Hart & Berendsen, 1981; Knapp, 1994). It is possible with the use of different combinations of these characters to separate the Telephiinae unequivocally from the Kalanchoeae. The *Telephium*-clade is only weakly supported in all chloroplast DNA phylogenies (Fig. 1; van Ham, 1994, this volume) and the alternative phylogeny resulting from Dollo parsimony analysis of the chloroplast DNA RFLPs (van Ham, 1994, this volume Fig. 3) emphasises its affinities with the Kalanchoeae. In the classification proposed here the Telephiinae bridge the gap between the African Kalanchoeae and the Northern Hemisphere Sedinae.

Subtribus Sedinae comprises about one half of the species and genera of the Crassulaceae. It contains the genus *Sedum* sensu stricto which appears to be paraphyletic in all analyses (Fig. 1; van Ham, 1994, this volume). Van Ham distinguished four clades within this well-defined lineage. In the chloroplast DNA RFLP analysis the strongly supported *Acre*-clade either was in a sister-group relationship with a monophyletic lineage containing the *Aeonium*-clade, *Leucosedum*-clade, and *Sempervivum*-clade (Dollo parsimony and Neighbor-joining) or in a monophyletic relationship with the *Aeonium*-clade and the *Leucosedum*- and *Sempervivum*-clade in a basal position. The monophyly of the *Acre*-clade is supported by phytochemical features such as the occurrence of alkaloids, which are a unique feature of this group (Stevens & al., 1995, this volume), and several morphological characters like the ornamentation of the seeds, which have a reticulate testa, except for *Sedum litoreum* Guss. ('t Hart & Berendsen, 1981; Knapp, 1994), the absence of glandular thrichomes, and the apparently primitive, free, spirally arranged, generally unequal sepals of *Sedum* species at the base of this clade ('t Hart & Koek-Noorman, 1989; van Ham, 1994). The taxa of the other three clades, on the other hand, always have costate (or bipapillate) seeds ('t Hart & Berendsen, 1981; Knapp, 1994) and basally fused sepals. Moreover they are often glandular pubescent. These two lineages of *Sedum* each gave rise to a great number of taxa which in Berger's (1930) classification belong to the subfamilies Echeverioideae, Sempervivoideae, and Cotyledonoideae (Eurasian and Mediterranean taxa). The *Acre*-clade is most diverse in Mexico and comprises, in addition to *Sedum*, the genera *Echeveria*, *Graptopetalum*, *Lenophyllum*, *Pachyphytum*, *Tacitus*, *Thompsonella*, and

Villadia (Thiede, this volume). The combined *Aeonium-*, *Leucosedum-*, an *Sempervivum*-clades, on the other hand, have a major centre in Eurasia and a secondary centre in Northern America. In addition to *Sedum* they comprise such Eurasian genera as *Aeonium*, *Aichryson*, *Monanthes*, and *Sempervivum* as well as the American genus *Dudleya*. In order to make these major evolutionary developments apparent taxonomically, the two lineages within *Sedum* have been classified as subgenera. Although at least the *Acre*-clade appears to be monophyletic the two subgenera are both paraphyletic. Consequently it is still extremely difficult or even almost impossible to separate *Sedum* from the other genera of the Sedinae.

The best solution of this taxonomic problem would be to divide *Sedum* into its monophyletic constituent taxa. However, this would result in a tremendous increase of usually very small or even monotypic, segregate genera (see for example Mes, this volume). Since, moreover, evolutionary relationships among the several hundreds of *Sedum* species are only partly known yet, only a fraction of these monophyletic groups can be indicated with certainty. Consequently, the untimely fragmentation of *Sedum* would merely result in unnecessary and unjustified changes of names as unfortunate attempts along this line by Grulich (1984) and Löve & Löve (1985a,b) have proven ('t Hart, 1991). For the Eurasian region the number of genera would increase to 30 or more (including several monotypic genera) most of which would be rather ill-defined in comparison to genera such as *Aeonium* or *Monanthes*. In my opinion this is not a realistic solution at present and for the sake of the stability of nomenclature I advocate the retention of an united, albeit paraphyletic, *Sedum*.

The latter alternative, however, raises the problem which criteria the segregate genera must meet in order to become recognized. Firstly, all segregate genera should be monophyletic. The constituent monophyletic taxa of a polyphyletic genus can either be sunk into *Sedum* or be given generic rank in their own right. A good example of the first solution is *Mucizonia*. The two or three species which are generally assigned to this small genus, which Berger (1930) classified in the subfamily Cotyledonoideae, belong to three different series (comparia) of *Sedum* and consequently *Mucizonia* has been merged with *Sedum* ('t Hart, 1991). An example of the second strategy is *Rosularia*. In Ohba's (1978) and Eggli's (1988) accounts of *Rosularia* this genus comprises at least 5 independent lineages and can most certainly not be upheld. In the account presented here the monophyletic taxa from South-Western Asia have been retained as genera, but the species of the Western Mediterranean region are again included in *Sedum* (Fig. 1; Mes & 't Hart 1994, Mes, this volume). Secondly, monotypic genera should be dropped if they merely contain a highly specialized form of relatively recent origin. A good example of this category is the genus *Telmissa*. The only species, *T. microcarpa* (Smith) Boiss. (*Sedum microcarpum* (Smith) Schönl.), is a morphologically very distinct, small annual which is adapted to the arid conditions of (semi-)desert habitats in the Near East, but phylogenetically it is closely related to annual *Sedum* taxa of the same region (van Ham, 1994; 't Hart & Alpinar, in press). Monotypic genera apparently representing ancient, isolated lineages or paleoendemics, on the other hand, may be retained until their evolutionary position has been elucidated. Examples of the latter group of monotypic genera are the African *Afrovivella*,

Hypagophytum, and *Perrierosedum*.

For two categories of monophyletic segregate genera it will be very difficult to decide whether they should either be retained or sunk into *Sedum*, viz. the morphologically highly distinct, small (2 to 3 species) segregate genera and the moderately large (4 to 15 species), morphologically ill-defined genera. *Pistorinia* with 2 or 3 species with their unique, brightly coloured, tubular flowers from the Western Mediterranean region is a good example of the first group. It is obviously monophyletic, but of relatively recent origin and closely related to *Sedum* of the same region (Fig. 1; van Ham, 1994, this volume). Because of its very distinct flowers it has been retained in the account presented here. *Prometheum*, on the other hand, is clearly an example of the second, more dubious, group of segregate genera. It comprises about 7, perennial or biennial, rosulate species with 5-merous flowers and basally connate petals. These features evolved independently in different groups of *Sedum* and if *Prometheum* is retained as a genus one may wonder why for example *Sedum* series *Cepaea*, which is also monophyletic and morphologically almost equally distinct, is not given the same taxonomic status. To solve such taxonomic incongruences the "consensus idea" will be of great help.

Key to the infrafamilial taxa of Crassulaceae.

1 Leaves opposite, usulally with hydathodes along the margins. Flowers haplostemonous, 5(3-9) -merous.

I. subfam. Crassuloideae

- Leaves alternate, opposite, or whorled, with a single apical hydathode. Flowers obdiplostemonous (if haplostemonous leaves never opposite), 5 (4-32) -merous.

II. subfamily Sedoideae. → 2

2 Leaves decussate (rarely alternate), flat (rarely terete), crenate or dentate (rarely entire). Flowers sympetalous, 4 or 5 -merous. Seeds costate.

A. tribus Kalanchoeae

- Leaves usually terete or semiterete, entire, alternate (if decussate, flowers neither sympetalous nor seeds costate). Flowers 5 (4-32) -merous. Petals usually free and spreading. Seeds costate, (multi)papillate or reticulate.

B. tribus Sedeae. → 3

3 Plants usually perennial. Stems often basally woody, forming tubers or rhizomes. Leaves usually flat, dentate or crenate, rarely decussate or terete or entire. Flowers 5-merous (if polymerous leaves neither flat nor opposite or whorled). Sepals always basally connate. Petals usully free (if connate seeds multipapillate). Seeds costate or multipapillate.

α subtribus Telephiinae

- Perennial or annual. Shoots usually herbaceous. Leaves often rosulate or subrosulate, terete or semiterete, entire, rarely flat, dentate-serrate, and/or decussate. Flowers 5 (4-32) -merous. Sepals basally connate or free. Petals usually free and stellate patent, sometimes connate. Seeds reticulate, papillate or costate.

β subtribe Sedinae

Conspectus

Crassulaceae De Candolle, in Lamarck & De Candolle, Fl. Franá. ed. 3, 4(1), 382. 1805 (nomen conservandum). Type: *Crassula* L.
Synonym: Sempervivae [Sempervivaceae] Jussieu, Gen. Pl., 307. 1789.

I Subfam. Crassuloideae Berger in Engler & Prantl, Nat. Pfl.-fam. ed. 2, 383. 1930. Type, *Crassula* L. (only element included).
Crassula L. in Spec. Pl., 282. 1753.
Tillaea L. in Spec. Pl., 128, 1753.
Note. Cosmpolitan genus with 20-40, highly specialized, often ephemeral, semi-aquatic species. Genetic divergence between this genus and its sistergoup *Crassula* is of the same order of magnitude as between the major lineages of subfamily Sedoideae (van Ham, 1994, this volume) and *Tillaea* certainly needs to be reinstated.

II Subfam. Sedoideae Berger in Engler & Prantl, Nat. Pfl.-fam. ed. 2, 384. 1930. Lectotype (designated here), *Sedum* L.
Synonyms: subfam. Cotyledonoideae Berger (1930); subfam. Echeverioideae Berger (1930); subfam. Kalanchoideae Berger (1930); subfam. Sempervivoideae Berger (1930); subfam. Dudleyoideae P.H. Thomson (1994).

A Tribus Kalanchoeae 't Hart comb. et stat. nov.
Basionym: subfam. Kalanchoideae Berger in Engler & Prantl, Nat. Pfl.-fam. ed. 2, 383. 1930. Lectotype (designated here), *Kalanchoe* Adan.
Adromischus Lem. in Jard. Fleur. 2. Misc., 58-59. 1852.
Cotyledon L. in Spec. Pl., 429. 1753.
Tylecodon Tölken, Bothalia 12, 378. 1978.
Kalanchoe Adan. in Fam. Pl. 2, 248. 1763.
Bryophyllum Salis. in Parad. Londin., t. 3, 1805.

B Tribus Sedeae 't Hart, trib. nov. Type, *Sedum* L.
Folia plerumque teretia vel semiteretia, integra, alterna (aut si decussata flores nunc nec sympetali nec semina costata). Flores 5 (4-32) -meri. Petala plerumque libera et stellato-patentia. Semina costata, (multi)papillata vel reticulata.

α Subtribus Telephinae 't Hart subtr. nov. Type, *Hylotelephium* H. Ohba
Plantae plerumque perennes. Caules basin saepe lignosi, rhizomata vel tubera formantes. Folia plerumque applanata, dentata vel crenata, raro decussata vel teretia vel integra. Flores 5-meri (aut si polymeri nunc folia nec applatana nec opposita vel verticillata). Sepala basin semper receptaculo adnata. Petala plerumque libera (aut si connata nunc semina nec papillosa). Semina costata vel multipapillosa.
Rhodiola L. in Spec. Pl., 1035. 1753.
Pseudosedum (Boiss.) Berger in Engler & Prantl, Nat. Pfl.-fam. ed. 2, 18a, 465. 1930.

Hylotelephium H. Ohba in Bot. Mag. (Tokyo) 90, 46-47. 1977.
Orostachys Fisch. ex Berger in Engler & Prantl, Nat. Pfl.-fam. ed. 2, 18a, 463. 1930.

Note. Meterostachys Nakai (in Bot. Mag. (Tokyo) 49, 74 and 210. 1935) and *Sinocrassula* Berger (in Engler & Prantl, Nat. Pfl.-fam. ed. 2, 18a, 462-463. 1930) are closely related to *Orostachys* and probably better treated as subgenera of the latter.

Phedimus Rafinesque in Amer. Monthly Mag. Crit. Rev. 1, 438. 1817. Lectotype, *Phedimus stellatus* (L.) Rafin. (= *Sedum stellatum* L.) ('t Hart in Fl. Medit. 1, 52. 1991).

Phedimus subgen. *Phedimus*

Synonyms: *Asterosedum* Grulich in Preslia 56, 38. 1984; *Spathulata* (A. Boriss.) Löve & Löve in Taxon 34, 163. 1985; *Sedum* subgen. *Spathulata* (A. Boriss.) H. Ohba in J. Fac Sci. Univ. Tokyo, Sec. 3, 4, 178. 1978; *Sedum* subsect. *Spathulata* A. Boriss. in Novit. Syst. Pl. Vasc. Acad. Sci. USSR 6: 117. 1969; *Sedum* series *Involucrata* Maxim. ex Berger in Nat. Pfl.-fam. ed. 2, 18a, 449. 1930; *Sedum* series *Propontica* Berger in Nat. Pfl.-fam. ed. 2, 18a, 450. 1930; *Sedum* series *Stoloniferum* Fröd. in Acta Hort. Gothob. 7, app., 49. 1932.

Species included: *Phedimus obtusifolius* (C.A. Meyer) 't Hart comb. nov. (Basionym, *Sedum obtusifolium* C.A. Meyer in Enum. Pl. Cauc., 150. 1831); *Phedimus spurius* (Bieb.) 't Hart comb. nov. (Basionym, *Sedum spurium* Bieb. in Fl. Taur. Cauc. 1, 352. 1808); *Phedimus stellatus* (L.) Rafin. in Amer. Monthly Mag. Crit. Rev. 1, 438. 1817; *Phedimus stevenianus* (Rouy & Camus) 't Hart comb. nov. (Basionym, *Sedum stevenianum* Rouy & Camus in Fl. Fr. 7, 94. 1901); *Phedimus stoloniferus* (Gmelin) 't Hart comb. nov. (Basionym, *Sedum stoloniferum* Gmelin in Reise 3, 311, tab. 35.1774).

Phedimus subgen. *Aizoon* (Koch) 't Hart comb. et stat. nov.

Basionym: *Sedum* sect. *Aizoon* Koch, Syn. Fl. Germ., 259. 1835. Type, *Sedum aizoon* L.

Synonyms: *Aizopsis* Grulich in Preslia 56, 35. 1984; *Sedum* subgen. *Aizoon* (Koch) H. Ohba in J. Fac. Sci. Univ. Tokyo, Sec. 3, 12, 179. 1978; *Sedum* series *Aizoonta* Maxim. in Bull. Acad. Sci. St.Petersb. 29, 143. 1883; *Sedum* series *Aizoon* (Koch) Fröd. in Acta Hort. Gothob. 6, app., 77. 1931.

Species included: *Phedimus* aizoon (L.) 't Hart comb. nov. (Basionym, *Sedum aizoon* L. in Sp. Pl. 1, 430. 1753); *Phedimus ellacombianus* (Praeger) 't Hart comb. nov. (Basionym, *Sedum ellacombianum* Praeger in J. Bot. 54, 41. 1917); *Phedimus floriferus* (Praeger) 't Hart comb. nov. (Basionym, *Sedum floriferum* Praeger in J. Bot. 56, 149. 1918); *Phedimus hybridus* (L.) 't Hart comb. nov. (Basionym, *Sedum hybridum* L. in Sp. Pl. 1, 431. 1753); *Phedimus kamtschaticus* (Fisch. & Meyer) 't Hart comb. nov. (Basionym, *Sedum kamtschaticum* Fisch. & Meyer in Ind. Sem. Hort. Petrop. 7, 54. 1841); *Phedimus litoralis* (Komarov) 't Hart comb. nov. (Basionym, *Sedum litorale* Komarov in Bull. Jard. Bot. Acad. Sc. USSR, 30, 201. 1932); *Phedimus maximowiczii* (Regel) 't Hart comb. nov. (Basionym, *Sedum maximowiczii* Regel in Gartenfl., 355, t. 528. 1866); *Phedimus middendorfianus* (Maxim.) 't Hart comb. nov. (Basionym, *Sedum middendorfianum* Maxim. Prim. Fl. Amur., 116. 1859); *Phedimus odontophyllus* (Fröd.) 't Hart comb. nov. (Basionym, *Sedum odontophyllum* Fröd. in Acta Hort. Gothob. 7, app., 117. 1932); *Phedimus selskanianus* (Regel & Maack) 't Hart comb. nov. (Basionym, *Sedum selskanianum* Regel & Maack in Tent. Fl. Ussur., 66. 1861); *Phedimus sichotensis* (Worosch.) 't Hart comb. nov. (Basionym, *Sedum sichotense* Worosch. in Bull. Princ. Bot. Gard. Acad. Sci. USSR, 40, 51. 1961); *Phedimus sikokianum* (Maxim.) 't Hart comb. nov.(Basionym, *Sedum sikokianum* Maxim. in Bull. Acad. Imp., 737. 1892; Diagn. Pl. Nov. Asiat. 8, 1. 1893); *Phedimustakesimensis* (Nakai) 't Hart comb. nov. (Basionym, *Sedum takesimense* Nakai in Rep. Veg. Dagelet Isl., 36. 1919); *Phedimus zokuriensis* (Nakai) 't Hart comb. nov. (Basionym, *Sedum zokuriense* Nakai in J. Jap. Bot. 15, 674. 1939).

Umbilicus DC. in Bull. Sci. Soc. Philom. Paris 3, 1. 1801.

Perrierosedum (Berger) H. Ohba in J. Fac. Sci. Univ. Tokyo, Bot. 12, 166. 1978.

Note. Because of the *Sedum*-like flower-structure it is tentatively placed in the Telephiinae, but it has also some affinity with the Kalanchoeae. *Perrierosedum* may be a paleoendemic marking the migration route of the Crassulaceae from their South-Eastern African homeland to the Northern Hemisphere.

Hypagophytum Berger in Engler & Prantl, Nat. Pfl.-fam. ed. 2, 18a, 467-468. 1930.

Note. In contrast to earlier classifications seed characters unequivocally place *Hypagophytum* in the Sedoideae. Although, its position within the subfamily is still uncertain it has provisionally been placed in the Telephiinae because of the vegetative characters.

β Subtribus Sedinae 't Hart, subtrib. nov. Type, *Sedum* L.

Plantae perennes vel annuae. Surculi plerumque herbacei. Folia saepe rosulata vel subrosulata, teretia vel semiteretia, integra, raro applanata dentato-serrata decussata. Flores 5 (4-32) -meri. Sepala basin receptaculo adnata vel libera. Petala plerumque libera et stellato-patentia, interdum connata. Semina reticulata, papillosa vel costata.

Sedum L. in Spec. Pl., 430. 1753.

Sedum subgen. *Gormania* (Britton) R.T. Clausen in Bull. Torr. Bot. Club 69, 28. 1942.

Usually glandular pubescent. Sepals basally connate, usually equally long. Testa costate or bipapillate.

Note. Most probably all other 10 genera of the *Aeonium*-, *Leucosedum*-, and *Sempervivum*-clades, i.e. *Aeonium*, *Afrovivella*, *Aichryson*, *Dudleya*, *Monanthes*, *Pistorinia*, *Prometheum*, *Rosularia*, *Sempervivella*, and *Sempervivum*, have a common ancestor with species of this subgenus.

Sedum subgen. *Sedum*

Usually glabrous or rarely with non-glandular trichomes. Sepals basally free or connate, often strongly unequal. Testa reticulate or reticulo-papillate (except *S. litoreum*).

Note. Most certainly the other 7 genera of the *Acre*-clade, i.e. *Echeveria*, *Graptopetalum*, *Lenophyllum*, *Pachyphytum*, *Tacitus*, *Thompsonella*, and *Villadia*, share a common ancestor with the species of this *Sedum* lineage.

β1 Combined *Leucosedum-*, *Aeonium-*, and *Sempervivum*-clades

Rosularia (DC.) Stapf in Curtis's Bot. Mag. 149, t. 8985. 1923.

Note. In Ohba's (1978) and Eggli's (1988) accounts the genus is polyphyletic combining at least 5 widely different, independent lineages of Sedinae. Here it is restricted to Eggli's *R.* sect. *Rosularia* and some of the species of *R.* sect. *Ornithogalopsis* Berger and *R.* sect. *Sempervivella* (Stapf) Jansson. It comprises the taxa with small chromosomes and a base number x=9 ('t Hart & Eggli, 1989). Its distribution area ranges from the Western Himalayas to the East Mediterranean with a centre of diversity in the Near East.

Species included: *R. adenotricha* (Wallich ex Edgew.) Jansson, *R. blepharophylla* Eggli, *R. davisii* Muirhead, *R. lymaitica* (Boiss. & Hausskn.) Berger, *R. glabra* (E. Regel & Winkler) Berger, *R. globulariifolia* (Fenzl) Berger, *R. hausknechtii* (Boiss. & Reuter) Berger, *R. lineata* (Boiss.) Berger, *R. modesta* (BornmÅller) Jansson, *R. pallidiflora* (Holmboe) Meikle, *R. platyphylla* (Schrenk) Berger, *R. radicosa* (Boiss. & Hohenacker) Eggli, *R. rosulata* (Edgew.) H. Ohba, *R. sempervivum* (Bieb.) Berger, *R. serrata* (L.) Berger.

Prometheum (Berger) H. Ohba in J. Fac. Sci. Univ. Tokyo, Ser. III,12, 168. 1978.

Type, *Prometheum sempervivoides* (Bieb.) H. Ohba. Basionym: *Sedum* sect. *Prometheum* Berger in Nat. Pfl.-fam. ed. 2, 18a, 459. 1930.

Synonyms: *Sedum* sect. *Sempervivoides* Boiss. in Fl. Orient. 2, 776. 1872; *Rosularia* sect. *Sempervivoides* (Boiss.) A. Boriss. in Komarov, Fl. USSR 9, 117. 1939. *Rosularia* sect. *Chrysanthae* Eggli in Bradleya 6, 32. 1988.

Note. Typical Anatolian element ranging eastwards from Central Greece (*P. tymphaeum*) to the Caucasus and Northern Iran. Differs from the previous genus by its large chromosomes and a base number x=7 ('t Hart & Eggli, 1989).

Species included: *Prometheum aizoon* (Fenzl) 't Hart comb. nov. (Basionym, *Umbilicus aizoon* Fenzl in Pugil. Pl. Nov. Syriae & Tauri Occ Prim., 15. 1842); *Prometheum chrysanthum* (Boiss.) 't Hart comb. nov. (Basionym, *Umbilicus chrysantha* Boiss. in Diagn. Pl. Orient. Nov. 2(10), 15. 1849); *Prometheum pilosum* (Bieb.) H. Ohba; *Prometheum rechingeri* (Jansson) 't Hart comb. nov. (Basionym, *Rosularia rechingeri* Jansson in Acta Horti Gothob. 28, 1882-4. 1966); *Prometheum serpentinicum* (Wedermann) 't Hart comb. nov. (Basionym, *Umbilicus serpentinicus* Wedermann in Notizbl. Bot. Gart. Mus. Berlin-D. 14, 370. 1939); *Prometheum sempervivoides* (Bieb.) H. Ohba; *Prometheum tymphaeum* (Quézel & Contandriopoulos) 't Hart comb. nov. (Basionym, *Sedum tymphaeum* Quézel & Contandriopoulos in Taxon 16, 240. 1967).

Sempervivella Stapf in Curtis's Bot. Mag. 49, tab. 8985. 1923.

Synonyms: *Rosularia* sect. *Sempervivella* (Stapf) Jansson in Fl. Iranica 72, 25. 1970.

Note. A small, heterogeneous group of very diverse Asian taxa of uncertain affinities, which may either belong to *Rosularia* sensu stricto or to one of the other lineages of Sedinae (Ohba, 1978; Eggli, 1988).

Species included: *Sempervivella mucronata* (Edgew.) Berger, *Sempervivella alba* (Edgew.) Stapf .

Sempervivum L. in Spec. Pl., 464. 1753.

Note. Jovibarba (DC.) Opiz (in Seznam, 54. 1852.) is very closely related to *Sempervivum* and most probably better treated as a subgenus of the latter.

Aeonium Webb & Berth. in Hist. Nat. Iles Canar. 3(2,1), 184. 1840.

Aichryson Webb & Berth. in Hist. Nat. Iles Canar. 3(2,1), 180. 1840.

Monanthes Haw. in Saxifr. Enum., 68. 1821.

Pistorinia DC. in Prodr. 3, 399. 1828.

Afrovivella Berger in Engler & Prantl, Nat. Pflanzenfam. ed. 2, 18a, 466-467. 1930.

Note. The affinities of this monotypic Ethiopian genus are still unknown. It has been included in *Cotyledon, Rosularia, Sempervivum,* and *Umbilicus* (Eggli, 1988). It may be a paleoendemic element and is better kept separate until its evolutionary relatioships have been determined.

Dudleya Britt. & Rose in Bull. New York Bot. Gard. 3(9), 12-13. 1903.

β2 *Acre*-clade

Villadia Rose in Bull. New York Bot. Gard. 3(9), 3. 1903.

Graptopetalum Rose in Contr. U. S. Nation. Herb. 13, 296. 1911.

Tacitus Moran in Cact. Suc. Mex. 19, 73, 75-84, 94. 1974.

Thompsonella Britton & Rose in Contr. US Nation. Herb. 12, 391. 1909.

Lenophyllum Rose in Smithsonian Misc. Coll. 47, 159. 1904.

Echeveria DC. in Prodr. 3, 401. 1828.

Pachyphytum Link, Klotzsch & Otto in Allg. Gartenzeit. 9, 9. 1841.

Acknowledgements

I would like to thank Ted Mes for his contribution to the analysis of the molecular data and Gea Zijlstra (Utrecht) and Larry M. Mitich (Davis) for their help with matters of nomenclature.

References

Berger, A., 1930: Crassulaceae. In Engler, A. & Prantl, K. (eds.), Die Natürlichen Pflanzenfamilien, ed. 2, **18a**, 352-483. Verlag Wilhelm Engelmann, Leipzig.

Capon, R.J., Macleod, J.K. & Oelrichs, P.B. 1986. Bryotoxins B and C, toxic bufadienolide orthoacetates from the flowers of *Bryophyllum tubiflorum* (Crassulaceae). *Austral. J. Chem.* **39**: 1711-1715.

Clausen, R.T. 1959. *Sedum* of the Trans Mexican Volcanic Belt. Ithaca, New York.

Clausen, R.T. 1975. *Sedum* of North America north of the Mexican Plateau. Ithaca, New York.

Eggli, U. 1988. A monographic study of the genus *Rosularia* (Crassulaceae). *Bradleya* **6**(suppl.): 1-119.

Grulich, V. 1984. Generic division of Sedoideae in Europe and adjacent regions. *Preslia* **56**: 29 54.

Ham, R.C.H.J. van, 1994. Phylogenetic implications of chloroplast DNA variation in the Crassulaceae. Ph.D. Thesis, Utrecht University.

Hart, H. 't. 1982. The white-flowered *Sedum* species. 1. Principles of a phylogenetic classification of the Sedoideae (Crassulaceae) and the position of the white-flowered *Sedum* species. *Proc. Kon. Ned. Akad. Wet., Ser. C* **85**: 663-675.

Hart, H. 't. 1984. The white-flowered European *Sedum* species. 3. The systematic position of *S. stellatum* L. *Proc. Kon. Ned. Akad. Wet., Ser. C* **87**: 401-411.
Hart, H. 't. 1991. Evolution and classification of the European *Sedum* species (Crassulaceae). *Flora Medit.* **1**: 31-61.
Hart, H. 't & Alpinar, K. Biosystematic studies in *Sedum* (Crassulaceae) of Turkey. (in press).
Hart, H. 't & Koek-Noorman, J. 1989. The origin of the woody Sedoideae (Crassulaceae). *Taxon* **38**: 535-544.
Jensen, L.C.W. 1966. Comparative anatomical studies in three subfamilies of the Crassulaceae. Ph.D. Thesis, University of Minnesota. University Microfilms Inc., Ann Arbor, Michigan.
Jensen, L.C.W. 1968. Primary stem vascular patterns in three subfamilies of the Crassulaceae. *Amer. J. Bot.* **55**: 553-563.
Knapp, U. 1994. Skulptur der Samenschale und Gliederung der Crassualecae. *Bot. Jahrb. Syst.* **116**: 157-187.
Löve, A. & Löve, D. 1985a. Crassulaceae. In Löve, A. (ed.), Chromosome number reports LXXXVI, 163-164. *Taxon* **34**: 159-164.
Löve, A. & Löve, D. 1985b. Crassulaceae. In Löve, A. (ed.), Chromosome number reports LXXXVII, 350-351. *Taxon* **34**: 346-351.
Mauritzon, J. 1933. Studien über die Embryologie der Familien Crassulaceae und Saxifragaceae. Hakand Ohlsons, Lund.
Mes, T.H.M. & Hart, H. 't. 1994. *Sedum surculosum* and *S. jaccardianum* (Crassulaceae) share a unique 70bp deletion in the chloroplast DNA trnL (UAA) - trnF (GAA) intergeneic spacer. *Pl. Syst. Evol.* **193**: 213-221.
Merxmüller, H., Friedrich, H.C. & Grau, J. 1971. Cytotaxonomische Untersuchungen zur Gattungsstruktur von *Crassula. Ann. Naturhist. Mus. Wien* **75**: 111-119.
Ohba, H. 1978. Generic and infrageneric classification of the Old World Sedoideae (Crassulaceae). *J. Fac. Sci., Univ. Tokyo, Sect. 3.* **12**: 139-198.
Pilon-Smits, L. 1992. Variation and evolution of crassulacean acid metabolism in *Sedum* and *Aeonium* (Crassulaceae). Ph.D. Thesis, Utrecht University.
Quimby, M.W. 1939. The floral morphology of the Crassulaceae. Ph.D. Thesis, Mississippi.
Stevens, J.F., Hart, H. 't, Ham, R.C.H.J. van, Elema, E.T., Ent, M.M.V.X. van den, Wildeboer, M. & Zwaving, J.H. 1995. Distribution of alkaloids and tannins in the Crassulaceae. *Biochem. Syst. Ecol.* **23**: 157-165.
Thomson, P.H. 1994. Dudleya and *Hasseanthus* handbook. Bonsall Publications, Bonsall, California.
Thorne, R.F. 1983. Proposed new realignments in the angiosperms. *Nordic J. Bot.* **3**: 75-117.
Uhl, C.H. 1948. Cytotaxonomic studies in the subfamilies Crassuloideae, Kalanchoideae, and Cotyledonoideae of the Crassulaceae. *Amer. J. Bot.* **35**: 695-706
Uhl, C.H. 1961. Some cytotaxonomic problems in the Crassulaceae. *Evolution* **15**: 375-383.
Uhl, C.H. 1963. Chromosomes and phylogeny of the Crassulaceae. *Cactus Succ. J. (US)* **35**: 80-84.
Uhl, C.H. 1992. Polyploidy, dysploidy, and chromosome pairing in *Echeveria* (Crassulaceae) and its hybrids. *Amer. J. Bot.* 79: 556-566.
Wagner, H., Fischer, M. & Lotter, H. 1985. Isolation and structure determination of daigremontianin, a novel bufadienolide from *Kalanchoe daigremontiana. Planta Medica* **1985**: 169-170.
Wassmer, A. 1955. Vergleichend-morphologische Untersuchungen an den Blüten der Crassulaceen. Ph.D. Thesis, Winterthur.
Willis, J.C. 1922. Age and area. Cambridge University Press, London.
Yamagisha, T., Haruna, M., Yan, X.Z., Chang, J.J. & Lee, K.H. 1989. Antitumor agents, 110. 1,2 bryophyllin B, a novel potent cytotoxic bufadienolide from *Bryophyllum pinnatum. J. Nat. Products* **52**: 1071-1079.

CHAPTER 11

Toward a consensus classification of the Crassulaceae

Urs Eggli, Henk 't Hart & Reto Nyffeler

Abstract

The classification of the family Crassulaceae has remained basically unchanged since Berger (1930). Following extensive discussions among specialists of the family, a consensus classification containing a number of deviations from the treatment by Berger is proposed. A number of traditionally recognized genera have been accepted provisionally only to point out the frequent lack of knowledge which prevents a critical evaluation of their position. All genera ever described for the family are enumerated in alphabetical sequence, including their synonymy, typification, distribution, etc., and with critical notes appended where necessary. More recent systematic treatments are indicated for the major groups.

Introduction

The Crassulaceae are a medium-sized family with an estimated 1500 species. It is distributed throughout the Northern hemisphere as well as in Southern Africa, with conspicuous centres of diversity in Mexico, South Africa and the European and temperate Asiatic mountains. While the family does not include any crop plants, a considerable number of species are important horticultural items (*Kalanchoe blossfeldiana* hybrids, amongst others), and many have found their way into amateur collections. Throughout the family's distribution, its species contribute importantly to the plant communities in semi-arid regions. Moreover, a limited number of taxa from this family have been extensively used in physiological studies, and the most important physiological adaptation in succulent plant metabolism the Crassulacean Acid Metabolism (CAM) is indeed named after this family.

The classification of the Crassulaceae has remained basically unchanged since the treatment by Berger (1930) for the second edition of Engler & Prantl's "Die natürlichen Pflanzenfamilien", which has its root partially in one of the earliest published infrafamilial classifications by De Candolle (1828a,b), partially (for the American taxa) in the treatment of the family by Britton & Rose (1905).

The only recent changes of note are various attempts to subdivide the impractically large genus *Sedum* (various segregates published by Löve & Löve, and by Holub). In comparison with other families of succulent plants, name changes have been relatively few in the past fifty years, and a surprisingly small number of newly described or revived older genera have gained much recognition.

Various recent studies, especially van Ham (1994), have made it increasingly clear that the present infrafamilial classification based on Berger (1930) is incompatible with biological reality. While it is impossible to resolve all questions shortly, clarity has been gained for some groups, and the time seems ripe for an attempt to renew the family classification. The consensus classification proposed below is a first step towards this goal.

The present contribution is the result of many discussions with specialists in various systematic or geographical groups of the family. These discussions culminated in a 'consensus discussion' (see below on the meaning of 'consensus') held at the Phoenix Congress (April 1992) of the International Organization for Succulent Plant Study (IOS) whose section 'Crassulaceae and other succulents' has acted as catalyst in the process. The classification which follows is the result of these discussions, and it includes ideas put forward by various specialists taking part in the consensus discussions. The final decisions, however, have been made by the present authors. The following have participated, either by attendance at Phoenix, or by mail: M. Cházaro Basañez (Guadalajara), U. Eggli (Zürich, ZSS), M.G. Gilbert (London), H. 't Hart (Utrecht), D.R. Hunt (Kew), M. Kimnach (California), J. Meyrán (Mèxico D.F.), R. Nyffeler (Zürich), N.P. Taylor (Kew), J. Thiede (Kiel). R. Moran (California) and C.H. Uhl (Ithaca) have further added many details when reviewing an earlier version of this paper.

Classificaton by Consensus - an alternative approach

'Classification by consensus' was probably first introduced by an ad hoc working party of the International Organization for Succulent Plant Study (IOS) in the first half of the 1980s to tackle the multitude of problems which beset the classification of the family Cactaceae. The consensus approach has greatly facilitated the exchange of ideas amongst various specialists and has, as a first result, produced something like a minimal set of accepted genera, subscribing to a rather broad generic concept (Hunt & Taylor, 1986). This classification was considerably modified in the following years, based on input from many workers throughout the world, resulting in a revised classification (Hunt & Taylor, 1990). The consensus classification thus produced is not entirely accepted unanimously by all participants of the discussions, but it can serve as a minimal base classification for all those purposes where the classification is merely used as a vehicle to convey other information (nature conservation, research in other domains than 'pure' systematics, review papers, basis for cataloguing herbaria, etc.). The consensus classification should represent an 'international standard' for the taxonomy and nomenclature (Hunt & Taylor, 1986) and at the same time represents a consensus of current opinion among the specialists involved (Hunt & Taylor, 1990).

The consensus approach itself will not solve any problems, but it has become obvious both in the case of the Cactaceae (Hunt & Taylor, 1986, 1990) and the Crassulaceae (this paper) that the on-going discussions quickly reveal all those groups where current knowledge is insufficient to finally decide on systematic position. If anything, the consensus identifies such 'hot spots' where future research is urgently needed. In the meantime, and until such results become available, the current consensus opinion may serve as a (provisional) unifying classification. As such, it may replace the multitude of digressing specialist's opinions most of which are based on extensive studies of a limited number of taxa and not intended as surveys of the whole family.

Thus, a consensus classification should be seen as one stage on the (long) way to a classification reflecting 'biological truth'. Hopefully, the consensus discussions will eventually converge onto a single accepted classification, if anything like 'biological truth' exists, all opinions eventually must converge onto it, and current 'dissensus' of opinion is primarily seen as the result of insufficient data and knowledge. Accordingly, the consensus classification should not censor any opinion based on sound data and reasoning. While it is impossible to reflect all opinions of all participants in the resulting classification (which can be interpreted as showing a majority view or trend of opinion), appropriate notes have been appended in a number of cases where substantial disagreement persisted.

It should be stressed that the 'consensus classification' which has resulted from the discussions in the past few years, and which is presented below, does not necessarily give a taxonomically uniform treatment of different generas within the family. It is already apparent that narrower generic concepts have been favoured by the majority of the contributors when more workers have studied a given group, while rather broad concepts have been adopted when little recent research was done, or when only a single or few scientists have been involved. This is very evident when the rather conservative treatment of *Crassula* is compared with the much more finely dissected classification of the North American Echeverioideae with many closely related genera still accepted. In a more balanced treatment of the family, almost all of these currently accepted genera (i.e. *Echeveria*, *Graptopetalum*, *Pachyphytum*, *Tacitus*, *Thompsonella*, and including *Sedum* sect. *Pachysedum*, but excluding *Dudleya* and probably also *Villadia*) would be united under *Echeveria*, to render a similar treatment as adopted for *Crassula*.

It is our hope that, keeping these limitations and shortcomings in mind, this 'consensus classification' may be of use to scientists and laymen alike, and particularly that it may help to facilitate and stimulate communication between all students of Crassulaceae.

The genera of the Crassulaceae

The following consensus classification of the family utilizes a number of typographical conventions to distinguish various cases. Generic names are given in alphabetical sequence. Names accepted for the sake of the consensus classification are given in **bold** print, whereas those placed into the synonymy of something else are printed in *italics.*

Genera accepted for this first stage of the consensus classification of the Crassulaceae fall into two groups which in part follow the conventions used for the Cactaceae consensus (Hunt & Taylor, 1990):

A. Names for genera with unanimous support or supported by a vast majority of the participants are printed in 10 point type (like this).

B. Names for genera with significant support, as well as those with completely unknown systematic affiliations, are printed in 9 point type (like this).

In addition, the abbreviation **Dist.** stands for Distribution and precedes a very rough outline of the geographical distribution of the genus concerned, while **LST** stands for Last Systematic Treatment and precedes information on more recent monographic (rarely floristic) studies of the genus.

References

Berger, A. 1930. Crassulaceae. In Engler, A. & Prantl, K. (eds.), Die natürlichen Pflanzenfamilein, ed. 2, **18a**, 352-485. Verlag Wihelm Engelmann, Leipzig,

Britton, N.L. & Rose, J.N. 1905. North American flora, **22**.

De Candolle, A.P. 1828a. Prodromus naturalis regni vegetalibis, **3**. Treuttel & Würtz, Paris.

De Candolle, A.P. 1828b. Mémoire sur la famille des Crassulacées. Treuttel & Würtz, Paris.

Ham, R.C.H.J. van. 1994. Phylogenetic implications of chloroplast DNA variation in the Crassulaceae. Ph.D. Thesis, Utrecht University.

Hunt, D.R. & Taylor, N.P. (eds.). 1986. The genera of the Cactaceae: towards a new consensus. *Bradleya* **4**: 65-78.

Hunt, D.R. & Taylor, N.P. (eds.). 1990. The genera of Cactaceae: progress towards consensus. *Bradleya* **8:** 85-107.

Consensus Classification

Adromischus Lemaire, Jard. Fleur. 2. Misc., 58-59. 1852. Lectotype: *Cotyledon hemisphaericus* Linnaeus (selected by Jacobsen, Kakt. and. Sukk. 17, 189. 1966). The spelling is often given as *Andromischus*, but this is incorrect; the name is based on Gr. '[h]adros', meaning 'squat', 'sturdy', and Gr. 'mischos', meaning 'pedicel', and refers to the short and stout pedicels. Species: 27. **Dist.**: Southern Africa (RSA, Namibia). **LST**: Tölken, H.R., Fl. South Afr. 14, 37-60. 1985.

Notes: A. phillipsiae (Marloth) von Poellnitz is aberrant in many characters and might warrant a subgenus of its own (E. van Jaarsveld, unpublished data). It possibly represents an ancient stabilized *Adromischus* x *Cotyledon* hybrid, but likewise shows some characteristics of *Tylecodon.*

Aeonium Webb & Berthelot, Hist. Nat. Iles Canar. 3(2,1), 184. 1840. Lectotype: *Sempervivum arboreum* Linnaeus (designated by Liu, Syst. *Aeonium*, 35. 1989). The lectotypification as given above is disputed by Heath (Calyx 2, 56-59. 1992; Calyx 3, 52. 1993) where it is stated that Pfeiffer (Nomenclator Bot. 1, 67. 1871-74) selected *Sempervivum smithii* Sims as type. This argument is unfounded, however, since Pfeiffer refers to "*Aeonium smithii* etc." and can consequently be ignored (G. Zijlstra, pers. comm.). Synonyms: *Aldasorea* Hort. ex Haage & Schmidt (1930); *Greenovia* Webb & Berthelot (1840); *Megalonium* (A. Berger) G. Kunkel (1980). Species: c. 35. **Dist.**: Macaronesia, Morocco, East Africa and Southern Arabia. **LST**: Liu, H.Y., Syst. *Aeonium*. Nation. Mus. Nat. Sci., Taichung (Taiwan). 1989. (*Aeonium* s.str., key).

Notes: The relationship between the Macaronesian genera *Aichryson*, *Aeonium*, and *Greenovia* seems to be rather close, with *Monanthes* standing considerably apart. While there is agreement that not enough evidence for separating *Greenovia* from *Aeonium* is available (the particular placentation is reported by J. Thiede to be an error of observations; and natural hybrids between the two are known), *Aichryson* differs cytologically and should be maintained at generic rank. The two groups *Monanthes* + *Aeonium* + *Aichryson* and *Sempervivum* + *Jovibarba* appear to represent different lineages, based on chloroplast DNA studies (see van Ham, Mes, and 't Hart, this volume). The two groups differ moreover in their axis anatomy; *Sempervivum* and *Jovibarba* have axes without secondary wood, while *Aeonium* and *Aichryson* have woody axes (J. Thiede).

Afrovivella A. Berger, in Engler & Prantl, Nat. Pflanzenfam. ed. 2, 18a, 466-467. 1930. Type: *Umbilicus semiensis* Gay ex A. Richard (typification by inference, only element included). →*Rosularia*. Species: 1. **Dist.**: Ethiopia (Semien Mts.).

Aichryson Webb & Berthelot, Hist. Nat. Iles Canar. 3(2,1), 180. 1840. Type: *Sempervivum dichotomum* De Candolle. Typification according to Bramwell (Bol. Inst. Nac. Invest. Agron. 28, 204. 1968) *Aichryson laxum* (Haworth) Bramwell (= *Aichryson dichotomum*). Synonym: *Macrobia* (Webb & Berthelot) G. Kunkel (1977). Species: c. 10. **Dist.**: Macaronesia (Morocco ?). **LST**: Bramwell, D., Bol. Inst. Nat. Invest. Agron. 28, 203-213. 1968. (synopsis only, key).

Notes: Molecular studies showed *Aichryson* to be the sister group of *Aeonium*, *Greenovia*, and *Monanthes* (see Mes, this volume). It links the former genera to the tender yellow-flowered North African species of *Sedum* with polymerous flowers. The genus should be retained, at least provisionally, on account of its morphology (in comparison with *Aeonium* and *Monanthes*), and its different basic chromosome number(s). Moreover, in contrast to the alleged intergeneric hybrid x *Aeonichryson* P.V. Heath (Calyx 2, 59. 1992), it appears to be reproductively isolated from *Aeonium* (H. 't Hart).

Aithales Webb & Berthelot, Hist. Nat. Iles Canar. 3(2,1), 179. 1840. Lectotype: *Sedum rubens* Linnaeus (designated by 't Hart, Fl. Medit. 1, 52. 1991). →*Sedum.*

Aizopsis Grulich, Preslia 56, 35-36. 1984. Type: *Sedum aizoon* Linnaeus. Based on *Sedum* sect. *Aizoon* Koch ex Schönland (in Engler & Prantl, Nat. Pflanzenfam. 3(2a), 30. 1891). →*Sedum.*

Aldasorea Hort. ex Haage & Schmidt, Gen. Cat., 203, 1930. Type: not designated. Cited from Index Kewensis. The name was first used (on herbarium sheets only) by Rose for the plant he later described as *Graptopetalum pachyphyllum* (see R. Moran, Cact. Succ. J. 62, 29. 1990). →*Aeonium.*

Altamiranoa Rose, Bull. New York Bot. Gard. 3(9), 31. 1903. Type: *Cotyledon batesii* Hemsley. →*Villadia.*

Amerosedum Löve & Löve, Taxon 34, 250. 1985. Type: *Sedum lanceolatum* Torrey. →*Sedum.*

Anacampseros Miller, Gard. Dict. Abr. ed. 4. 1754. Type: not designated. *Nomen rejiciendum versus Anacampseros* Linnaeus (1758), (*nom. cons.*). Often attributed to Haworth (Syn. Pl. Succ., 111. 1812). The name dated from pre-Linnean times and was already used by Bauhin or Tournefort. →*Sedum.*

Asterosedum Grulich, Preslia 56, 38-39. 1984. Type: *Sedum stellatum* Linnaeus. →*Sedum.*

Baumgartenia Tratinnick, Auswahl vorzügl. schöner Gartenpfl. 1, 110-112, pl. 1. 1821. Type: not designated. *Non Baumgartenia* Sprengel (1817) [= *Borya* Labillardiäre, Liliaceae]. →*Bryophyllum.*

Breitungia Löve & Löve, Taxon 34, 350. 1985. Type: *Sedum oreganum* Nuttall. Based on *Sedum* subgen. *Gormania* sect. *Oreganica* R.T. Clausen (Bull. Torrey. Bot. Club 69, 28. 1942). →*Sedum.*

Bryophyllum Salisbury, Parad. Londin., t. 3, 1805. Type: *Bryophyllum calycinum* Salisbury. Synonyms: *Baumgartenia* Tratinnick (1821); *Crassuvia* Commerson ex Lamarck (1786); *Kitchingia* Baker (1881); *Physocalycium* Vest (1820). Species: approx. 35. **Dist.:** Madagascar.

Notes: 2 species now widely distributed as pantropical weeds. The natural occurrence of the genus on mainland South Africa is open to debate. The split from *Kalanchoe* s.str. is in continued debate, and both genera need further study. The two genera are here separated provisionally and on practical grounds, more or less repeating the treatment given to *Cotyledon* / *Tylecodon* (for further arguments see Lauzac-Marchal, Compt. Rend. Acad. Sci. Paris ser. D, 278. 2505-2508. 1974). This very possibly leaves *Kalanchoe* paraphyletic.

Bulliarda De Candolle, Bull. Soc. Philom. 3 (no. 49), 1. 1801. Type: *Tillaea aquatica* Linnaeus (typification according to Britton, Bull. New York Bot. Gard. 3(9), 1. 1903). *Bulliarda* Necker (1790) [Annonaceae] is sometimes regarded as an earlier homonym, but is invalid (Art. 20, Ex. 11). →*Tillaea.*

Byrnesia Rose, Addisonia 7, 37. 1922. Type: *Byrnesia weinbergii* Rose (typification by inference, only element included). →*Graptopetalum.*

Calanchoe Persoon, Syn. Pl. 1, 445. 1805. Type: not designated. An orthographic variant of *Kalanchoe.* →*Kalanchoe.*

Cepaea Fabricius, Enum., 147. 1759. Type: *Sedum cepaea* Linnaeus (typification by inference, only element included). Also attributed to Fourreau (Ann. Soc. Linn. Lyon, ser. nov. 16, 384. 1868). The name dates back to pre-Linnean times and was probably first used by Caesalpinus (see Fourreau, l.c.). →*Sedum.*

Chamaerhodiola Nakai, in Nakai & Kitagawa, Rep. 1. Sci. Exped. Manch., sect. 4, 1, 27. 1934. Type: *Sedum quadrifidum* Pallas. Probably based on *Rhodiola* group *Chamae-Rhodiola* Fischer & Meyer (1841), which is included as synonym. →*Rhodiola.*

Chetyson Rafinesque, Ann. Gen. Sci. Phys. 7, 387. 1821. Type: *Chetyson sagittatum* Rafinesque (Typification according to ING.). →*Sedum.*

Chiastophyllum (Ledebour) Stapf ex A. Berger, in Engler & Prantl, Nat. Pflanzenfam. ed. 2, 18a, 418. 1930. Type: *Cotyledon oppositifolium* Ledebour. Based on *Umbilicus* sect. *Chiastophyllum* Ledebour (Fl. Ross. 2, 176. 1845). Also attributed to Stapf (in Index Londinensis 2, 176, 316. 1930). →*Umbilicus.* Species: 1. **Dist.:** Western Caucasus (USSR).

Clausenellia Löve & Löve, Taxon 34, 350. 1985. Type: *Sedum ternatum* Michaux. Based on *Sedum* sect. *Ternata* A. Berger (in Engler & Prantl, Nat. Pflanzenfam. ed. 2., 18a, 450. 1930). →*Sedum.*

Clementsia Rose, Bull. New York Bot. Gard. 3(9), 3. 1903. Type: *Sedum rhodanthum* A. Gray. →*Rhodiola.*

Cockerellia (Clausen & Uhl) Löve & Löve, Taxon 34, 351. 1985. Type: *Sedum cockerellii* Britton. Based on *Sedum* sect. *Cockerellia* Clausen & Uhl (Brittonia 5, 34-35. 1947). →*Sedum.*

Combesia A. Richard, Tent. Fl. Abyss. 1, 307. 1848. Type: *Combesia abyssinica* A. Richard (typification by inference, only element included). Publication year given here according to ING and TL-2, often erroneously cited as 1847. →*Tillaea.*

Notes: Tentatively referred here (H.'t Hart).

Congdonia Jepson, Man. Fl. Pl. Calif., 450. 1925. Type: *Sedum pinetorum* Brandegee. *Non Congdonia* J. Müller (1876) [Rubiaceae]. →*Sedum.*

Corynephyllum Rose, North Amer. Fl. 22, 28. 1905. Type: *Corynephyllum viride* Rose (typification by inference, only taxon included). →*Sedum.*

Cotyledon Linnaeus, Spec. Pl., 429. 1753. Lectotype: *Cotyledon orbiculata* Linnaeus (proposed for conservation by Tölken in Jarvis, Taxon 41, 561. 1992). Green (in

Hitchcock, Prop. Brit. Bot. 1929), proposed *Cotyledon hemisphaerica* Linnaeus, which belongs to *Adromischus*, however. The newly proposed lectotype is based on Hermann (Hort. Lugd. Bat. Cat., t. 551, 1687). Heath (Calyx 5, 22-26. 1994) reviewed the history of the names involved (incl. *Umbilicus*). Species: 9. **Dist.:** Southern & tropical East Africa, South-West Arabia. **LST:** Tölken, H.R., Fl. South Afr. 14, 3-17. 1985. (floristic-monographic treatment).

Cotyliphyllum Link, Handbuch 2, 23. 1831. Type: not designated. →*Umbilicus.*

Courantia Lemaire, Jardin Fleur. 1, Misc., 92. 1851. Type: *Courantia echeverioides* Lemaire (typification by inference, only element included). *Non Courantia* Schultz-Bip. (1844) [Papilionaceae]. Rose (Bull. New York Bot. Gard. 3(9), 10-11. 1903) lists *Echeveria rosea* Lindley as type of the genus and gives the Lemairean type as synonym. →*Echeveria.*

Crassula Linnaeus, Spec. Pl., 282. 1753. Type: *Crassula perfoliata* Linnaeus (designated by Hitchcock, Prop. Brit. Bot., 156. 1929. et errata). See also Heath (Calyx 3, 78. 1993). Synonyms: *Creusa* P.V. Heath (1993); *Curtogyne* Haworth (1821); *Danielia* (De Candolle) Lemaire (1869); *Dietrichia* Trattinnick (1812); *Dinacria* Harvey (1862); *Disporocarpa* (C.A. Meyer) A. Richard (1847); *Globulea* Haworth (1821); *Gomara* Adanson (1763); *Grammanthes* De Candolle (1828); *Kalosanthes* Haworth (1821); *Larochea* Persoon (1805); *Pagella* Schönland (1921); *Petrogeton* Ecklon & Zeyher (1837); *Purgosea* Haworth (1828); *Pyrgosea* Sweet ex Ecklon & Zeyher (1837); *Rhopalota* N.E. Brown (1931); *Rochea* De Candolle (1802); *Sarcolipes* Ecklon & Zeyher (1837); *Septas* Linnaeus (1761); *Septimia* P.V. Heath (1993); *Sphaeritis* Ecklon & Zeyher (1837); *Tetraphyle* Ecklon & Zeyher (1837); *Thisantha* Ecklon & Zeyher (1837); *Tillea* Sanguinetti (1852); *Toelkenia* P.V. Heath (1993); *Turgosea* Haworth (1821); *Vauanthes* Haworth (1821). Species: approx. 200. **Dist.:** mainly Southern Africa. **LST:** Tölken, H.R., Fl. South Afr. 14, 75-229. 1985. (South African taxa only).

Notes: This genus rivals *Sedum* for the number of generic synonyms, and the number gives a good idea of the diversity within the genus in its broadest sense. *Pagella* is often separated on account of the unique morphology / anatomy (see Jäger-Zürn, Trop. subtrop. Pflanzenwelt 70. 1989). *Tillaea* is here recognized at generic rank but is frequently united with *Crassula.*

Crassularia Hochstetter ex Schweinfurth, Beitr. Fl. éthiop., 80, in synon. 1867. Type: *Crassularia sediformis* Hochstetter ms. The name *Crassularia sediformis* (Hochstetter in Buchinger's Verz. Nr. 126) is only mentioned in the synonymy of *Crassula sediformis* by Schweinfurth when he published this taxon as new. →*Sedum.*

Notes: *Crassula sediformis* Schweinfurth is treated as *Sedum crassularia* R. Hamet by Gilbert (Bradleya 3, 48. 1985).

Crassuvia Commerson ex Lamarck, Encycl. 2, 141. 1786. Type: not designated. *Crassouvia* in De Candolle (Prodr. 3, 395. 1828) is a spelling error. →*Bryophyllum.*

Cremnophila Rose, North Amer. Fl. 22, 56. 1905. Type: *Sedum nutans* Rose. Species: 2. **Dist.:** Mexico (Mexico, Morelos). **LST:** Moran, R., Cact. Succ. J. (US) 50, 139-146. 1978. (ills., key).

Notes: The two species placed here by Moran (l.c.) are intermediate in many respects between *Echeveria* and *Sedum*. The genus was not retained by the consensus participants (*C. nutans* = *Sedum*; *C. linguifolia* = *Echeveria*) but is here provisionally kept as a matter of convenience; it shares many characters with *Pachyphytum*, including base chromosome number (x=33) and is moreover similar to some members of *Sedum* sect. *Pachysedum*. The characteristic pendulous inflorescence is diagnostic.

Creusa P.V. Heath, Calyx 3, 103. 1993. Type: *Crassula acutifolia* Lamarck. Based on *Crassula* group *Acutifolia* Schönland (Trans. Roy. Soc. South Africa 17, 167. 1929). →*Crassula.*

Curtogyne Haworth, Rev. Pl. Succ., 8, 1821. Lectotype: *Crassula undata* Haworth. (according to lectotypification of *Crassula* sect. *Curtogyne* by Tölken, Contrib. Bolus. Herb. 8, 300-301. 1977). →*Crassula.*

Danielia (De Candolle) Lemaire, Pl. Grasses, 94, 1869. Type: not designated. Based on *Rochea* sect. *Franciscea* De Candolle (Prodr. 3, 393. 1828). *Non Danielia* Correa de Mello ex Verlot (1868) [Bignoniaceae], *nec Daniellia* Bennett (1854) [Caesalpiniaceae], *nec Danielia* Skvortzov (1967) [Hexamitophyceae]. →*Crassula.*

Dasystemon De Candolle, Prodr. 3, 382. 1828. Type: *Dasystemon calycinum* De Candolle (typification by inference, only element included). →*Tillaea.*
Notes: Tentatively referred here.

Diamorpha Nuttall, Gen. North Amer. Pl. 1, 293. 1818. Type: *Sedum pusillum* Michaux (= *Diamorpha smallii* Britton; the conserved type is a Nuttall-specimen in PH.). *Non Diamorpha* Martius ex E.M. Fries (1832), (Fungi). Species: 1. **Dist.:** USA (granite outcrop endemic).
Notes: See Wilbur (Sida 13, 1-16. 1988) on the complicated nomenclature and taxonomy of this genus. The same author has proposed the name for conservation, because of problems in the application of the type species (Taxon 33, 116-118. 1984). The genus is probably an old specialized derivation of *Sedum* and could be united with that (H. 't Hart).

Dietrichia Trattinnick, Arch. Gewächskunde 1812, 449. 1812. Type: not designated. *Non Dietrichia* Giseke (1792) [= *Zingiber*, Zingiberaceae]. →*Crassula.*

Dinacria Harvey, in Harvey & Sonder, Fl. Cap. 2, 330. 1862. Type: *Grammanthes filiformis* Ecklon & Zeyher (typification by inference, only element included). →*Crassula.*

Diopogon Jordan & Fourreau, Brev. Pl. Nov. 2, 46. 1868. Type: *Sempervivum hirtum* Linnaeus. Based on *Sempervivum* sect. *Jovisbarba* Koch (Syn. Fl. Germ. Helv., ed. 2, 290. 1843), (= *Sempervivum* sect. *Jovibarba* [sic] Koch non De Candolle). →*Jovibarba.*

Diotostemon Salm-Dyck, Allg. Gartenzeit. 22, 265. 1854. Type: *Diotostemon hookeri* (typificaction by inference, only element included). →*Pachyphytum.*

Disporocarpa (C.A. Meyer) A. Richard, Tent. Fl. Abyss. 1, 307, in synon. 1847. Type: not designated. Based on *Crassula* subgen. *Disporocarpa* C.A. Meyer (Ind. Hort. Petr. 8, 56. 1841). The name is mentioned in the synonymy of *Combesia.* →*Crassula.*

Dudleya Britton & Rose, Bull. New York Bot. Gard. 3(9), 12-13. 1903. Type: *Echeveria lanceolata* Nuttall. Synonyms: *Hasseanthus* Rose (1903); *Stylophyllum* Britton & Rose (1903). Species: c. 40.

Dist.: Western USA, North-West Mexico. **LST:** Moran, R., in H. Jacobsen, Sukkulentenlexikon. 1970.

Notes: Despite the work by Moran, no modern monograph of the genus exists, and the number of described species seems exaggerated, taking into account the apparent variability and the occurrence of natural hybrids. In respect to *Echeveria*, the genus is very probably of independent origin from *Sedum*, and despite the availability of the hybrid name x *Dudleveria*, no confirmed hybrids between species of the two genera are known (C.H. Uhl). In addition to morphological and cytological differences, *Dudleya* seems to differ also in the structure of its epidermal waxes (J. Thiede). Thomson (*Dudleya* and *Hasseanthus* Handbook. Publ. by the author, Bonsall (USA). 1993) proposed the subfamily Dudleyoideae (nom. inval.) for the two genera *Dudleya* and *Hasseanthus*.

Echeveria De Candolle, Prodr. 3, 401. 1828. Lectotype: *Cotyledon coccinea* Cavanilles (designated by Rose, Bull. New York Bot. Gard. 3, 5. 1903). Synonyms: *Courantia* Lemaire (1851); *Oliveranthus* Rose (1905); *Oliverella* Rose (1903); *Urbinia* Britton & Rose (1903). Species: c. 130. **Dist.:** mostly Mexico, but some spp. from USA to Argentina. **LST:** Walther, E., Echeveria. California Acad. Sci., San Francisco. 1972.

Notes: The genus is mainly accepted by convenience and the apparently close relationships with *Sedum* (especially as to such probably aberrant taxa as *Echeveria sedoides* or *Sedum suaveolens*) need to be further evaluated (the latter placed in *Graptopetalum* by Clausen). *Cremnophila* presents further similar problems. The classification into sections by Walther (l.c.) is in need of revision. The systematics of a number of species groups is not resolved, and many taxa have been described since his publication. A revised sectional classification by Moran was included in Jacobsen (Sukkulentenlexikon, 163-165. 1970). The question can be raised whether all North American Crassulaceae with lateral inflorescences (*Echeveria*, *Graptopetalum*, *Pachyphytum*, *Tacitus*, *Sedum* sect. *Pachysedum*) should be united into an amplified genus *Echeveria*, at least insofar as they are reported to be interfertile (J. Thiede, J. Meyrán).

Enchylus Erhart, Beitr. 4, 147. 1789. Type: *Sedum annuum* Linnaeus (typification by inference, only element included). The name is only mentioned in a list and is without description, and therefore considered invalid. →*Sedum*.

Etiosedum Löve & Löve, Taxon 34, 163. 1985. Type: *Sedum annuum* Linnaeus. →*Sedum*.

Globulea Haworth, Syn. Pl. Succ. 60, 1821. Type: not designated. →*Crassula*.

Gomara Adanson, Fam. 2, 248. 1763. Lectotype: *Crassula pellucida* Linnaeus (designated by Heath, Calyx 3, 72. 1993). *Non Gomara* Ruiz & Pavón (1794) [Scrophulariaceae]. *Gomaria* Sprengel (1831) is an orthographic variant →*Crassula*.

Gormania Britton, Bull. New York Bot. Gard. 3(9), 29. 1903. Type: *Cotyledon oregonensis* S. Watson (= *Gormania watsonii* N.L. Britton, *nom. illeg.*). →*Sedum*.

Grammanthes De Candolle, Prodr. 3, 392. 1828. Type: not designated. Illegitimate replacement name for *Vauanthes* Haworth (1821). →*Crassula*.

Graptopetalum Rose, Contr. U. S. Nation. Herb. 13, 296. 1911. Type: *Graptopetalum pusillum*

Rose (typification by inference, only element included). Synonym: *Byrnesia* Rose (1922). Species: c. 12. **Dist.:** USA (Arizona) and Mexico (to Oaxaca). **LST:** Revision almost completed by R. Moran.

Notes: For the time being it would be premature and impractical to sink this genus into *Sedum*, but the similarities are obvious (e.g. *Sedum allantoides* with similar petal markings, although these might by the result of convergency (C.H. Uhl). The single species of *Tacitus* (here provisionally accepted on account of suggestions from a strong minority of the participants) differs only in a well-defined different pollination syndrome from the 'true' *Graptopetalum* species. Clausen places two aberrant *Sedum* species (*S. craigii, S. suaveolens*) here, which gives another indication of extremely reticulate classificatory ideas present.

Greenovia Webb & Berthelot, Hist. Nat. Iles Canar. 3(2,1), 198. 1840. Type: *Sempervivum aureum* C. Smith (typification according to ING). Spelling changed to *Greenoughia* by Heath (Calyx 2, 59. 1992), but to be treated as intentional latinization. →*Aeonium*. Species: 4. **Dist.:** Canary Islands. **LST:** Praeger, R.L., Account of the *Sempervivum* group. Rot. Hort. Soc., London. 1932.

Notes: The genus is so closely related to *Aeonium* that it cannot be maintained separate; see comment under the former (see Mes, this volume).

Hasseanthus Rose, Bull. New York Bot. Gard. 3(9), 37. 1903. Type: *Sedum variegatum* S. Watson. →*Dudleya*.

Helladia M. Král, Preslia 59, 309. 1987. Type: *Sedum creticum* Presl. →*Sedum*.

Helophytum Ecklon & Zeyher, Enum. Pl. Afr., 288, 1836. Lectotype: *Crassula natans* (Thunberg) Ecklon & Zeyher (according to lectotypification of *Crassula* sect. *Helophytum* by Tölken, Contrib. Bolus. Herb. 8, 84-86. 1977). →*Tillaea*.

Hjaltalinia Löve & Löve, Taxon 34, 163. 1985. Type: *Sedum villosum* Linnaeus. →*Sedum*.

Hylotelephium H. Ohba, Bot. Mag. (Tokyo) 90, 46-47. 1977. Type: *Sedum telephium* Linnaeus. The validity of this name depends on the lectotypification of *Sedum* (see there). Synonym: *Telephium* J. Hill (1756). Species: c. 25. **Dist.:** Europe, Inner & Eastern Asia, North America.

Notes: It is questionable whether this genus can be upheld or whether it is preferrable to treat it as subgenus of *Sedum* (see Clausen, *Sedum* North America. 1975). While the genus certainly represents a distinct group showing many primitive characters (tendency for thickened roots; flat leaves; inflorescence architecture, it is difficult to separate from *Sedum*. DNA studies link it with 'flat-leaved Asiatic *Sedum* species' (H. 't Hart).

Hypagophytum A. Berger, in Engler & Prantl, Nat. Pflanzenfam. ed. 2, 18a, 467-468. 1930. Type: *Sempervivum abyssinicum* Hochstetter (typification by inference, only element included.). Species: 1. **Dist.:** Ethiopia. **LST:** Gilbert M.G., in Fl. Ethiopia 3, 10-11. 1989; Opera Bot. 121, 47-50. 1993. (ill.).

Notes: The single species placed here is aberrant in most characters and has previously been classified as *Sempervivum*, *Sedum*, *Aichryson*, or *Crassula*. If material should become available in cultivation, a detailed morphological and anatomical study may resolve its position. It is notable that most of its aberrant characters are also to be found in one or another species of *Crassula* which would favour inclusion in *Crassula* s.l. (M.G. Gilbert; J. Thiede).

Jovibarba (De Candolle) Opiz, Seznam, 54. 1852. Lectotype: *Sempervivum hirtum* Linnaeus (designated by Borissova, Novost. Sist. Vyssh. Rast. 6, 113. 1969). Based on *Sempervivum* sect. *Jovibarba* De Candolle (Prodr. 3, 413, 1828). Synonym: *Diopogon* Jordan & Fourreau (1868). Species: c. 2. **Dist.:** Mountains of East & South-Eastern Europe. **LST:** Parnell, J., Acta Bot. Hung. 34, 219-222. 1988. (synopsis, 2 species accepted only).

Notes: The spelling (often corrected to *Jovisbarba*) and validity of the name (vs. *Diopogon*) have been frequently disputed, but careful consultation of the original sources have shown beyond doubt that it is to be accepted (see Eggli & Nyffeler, Bot. Helv. 102, 171-173. 1992). The relationships of this genus with *Sempervivum* need to be evaluated more closely, and the question can be raised whether *Jovibarba* as presently defined might not be polyphyletic. Despite the agreement to sink *Jovibarba* into *Sempervivum*, the genus is here kept separate to draw attention to the cytological (*Jovibarba* with n=19, *Sempervivum* with n=16, 17, 18, 19; Uhl, Amer. J. Bot. 48, 114-123. 1961; ZÇsiger, Ber. Schweiz. Bot. Ges. 71, 113-117. 1961) and pollen morphological differences (Parnell, Kew. Bull. 46, 733-738. 1991). Moreover, there are no confirmed reports about hybrids between the two genera.

Kalanchoe Adanson, Fam. Pl. 2, 248. 1763. Lectotype: *Cotyledon laciniata* Linnaeus (lectotypification according to Tölken, Fl. South Afr. 14, 61. 1985). Synonyms: *Calanchoe* Persoon (1805); *Kalenchoe* Haworth (1812); *Verea* Willdenow (1799); *Vereia* H. Andrews (1798). Species: approx. 50. **Dist.:** Africa (especially South and Eeast), Arabia, Madagascar, Asia.

Notes: See *Bryophyllum* for a comment on the systematics of the group.

Kalenchoe Haworth, Syn. Pl. Succ., 109. 1812. Type: not designated. An orthographic variant for *Kalanchoe*. →*Kalanchoe.*

Kalosanthes Haworth, Rev. Pl. Succ., 6. 1821. Type: *Crassula jasminea* Haworth ex Sims. →*Crassula.*

Kirpicznikovia Löve & Löve, Bot. Not. 128: 515, 1976. Type: *Sedum quadrifidum* Pallas. Based on *Rhodiola* sect. *Chamae-Rhodiola* (Fischer & Meyer) A. Borissova (Novost. Syst. Vyssh. Rast. 6. 1969). →*Rhodiola.*

Notes: Only one of seven putative species of this segregate has been formally combined by Löve & Löve.

Kitchingia Baker, J. Linn. Soc. Bot. 18, 268. 1881. Type: not designated.→ *Bryophyllum.*

Larochea Persoon, Syn. Pl., 37. 1805. Type: not designated. Given as a homotypic synonym of *Rochea* by ING. →*Crassula.*

Lenophyllum Rose, Smithsonian Misc. Coll. 47, 159. 1904. Type: *Sedum guttatum* Rose. Species: 7. Dist.: USA (Southern Texas), North-Eastern Mexico. **LST:** Moran, R., Haseltonia 2, 1-19. 1994. (ills., key).

Notes: This is a weakly defined and poorly understood group with obvious similarities to *Sedum.*

Leucosedum Fourreau, Ann. Soc. Linn. Lyon, ser. nov. 16, 384. 1868. Lectotype: *Sedum album* Linnaeus (designated by H. Ohba, J. Fac. Sci. Univ. Tokyo, Bot. 12, 176. 1978). →*Sedum.*

Notes: Nom. inval. (Art. 32.1); published to include *Sedum album*, *S. dasyphyllum*, and *S.*

hirsutum (see Eggli, Bradleya 10, 83. 1992).

Macrobia (Webb & Berthelot) G. Kunkel, Cuad. Bot. Canar. 28, 35-37. 1977. Type: *Sempervivum tortuosum* Aiton. Based on *Aeonium* sect. *Macrobia* Webb & Berthelot (Hist. Nat. Iles Canar. 3(2,1), 183. 1840). → *Aichryson.*

Macrosepalum Regel & Schmalhausen, in Regel, Descr. Pl. Nov. Fedtschenko, 25, 1882. Type: *Macrosepalum turkestanicum* Regel & Schmalhausen. Typification according to ING. 't Hart (Fl. Medit. 1, 52. 1991) cites *Sedum aetnense* as type. →*Sedum.*

Megalonium (A. Berger) G. Kunkel, Kanar. Inseln und ihre Pflanzenwelt, 158. 1980. Type: *Sempervivum nobile* Praeger. Based on *Aeonium* sect. *Megalonium* A. Berger (in Engler & Prantl, Nat. Pflanzenfam. ed. 2, 18a, 429. 1930). →*Aeonium.*

Mesanchum Dulac, Fl. Hautes-Pyr., 320. 1867. Type: not designated. A homotypic synonym of *Tillaea* according to ING. →*Tillaea.*

Meterostachys Nakai, Bot. Mag. (Tokyo) 49, 74 [in adnot.], 210 [erratum]. 1935. Type: *Cotyledon sikokiana* Makino. *Merostachys* Nakai (l.c., 210) is a spelling error. Species: 1. **Dist.:** Southern Japan, Korea. **LST:** Moran, R., Cact. Succ. J. (US) 44, 262-273. 1972.

Notes: This monotypic genus might be related to *Orostachys* and the more *Sedum*-like taxa of *Rhodiola* sect. *Primuloides* (H. 't Hart). The gender of the generic name is to be treated as feminine (see *Orostachys*).

Monanthella A. Berger, in Engler & Prantl, Nat. Pflanzenfam. ed. 2, 18a, 446. 1930. Type: *Sedum jaccardianum* Maire & Wilczek. Nom. inval. (Art. 34.1). →*Sedum.*

Notes: Placed in the synonymy of *Rosularia* by Eggli (Bradleya 6, suppl. 1988), but on account of recently investigated living material, the taxon is placed in the genus *Sedum* (Mes, this volume).

Monanthes Haworth, Saxifr. Enum., 68. 1821. Type: *Sempervivum monanthes* Aiton (= *Monanthes polyphylla* Haworth; typification by inference, only element included). Synonym: *Petrophyes* Webb & Berthelot (1841). Species: 10. **Dist.:** Macaronesia. **LST:** Nyffeler, R., Bradleya 10, 49-82. 1992. (ills., key).

Mucizonia (De Candolle) A. Berger, in Engler & Prantl, Nat. Pflanzenfam. ed. 2, 18a, 419-420. 1930. Type: *Umbilicus hispidus* De Candolle (typification by inference, only element included). Based on *Umbilicus* sect. *Mucizonia* De Candolle (Prodr. 3, 399. 1828). Species: 3. **Dist.:** Iberian Peninsula, North Africa. → *Sedum*

Notes: In the present circumscription polyphyletic; each of the three species is genetically related to a different series (comparium) of *Sedum* sect. *Sedum* (H. 't Hart).

Oliveranthus Rose, North Amer. Fl. 22, 27. 1905. Type: *Oliverella elegans* Rose. *Nom. nov. pro Oliverella* Rose (1903) *non* van Tieghem (1895). →*Echeveria.*

Oliverella Rose, Bull. New York Bot. Gard. 3(9), 2. 1903. Type: *Oliverella elegans* Rose (typification by inference, only element included). *Non Oliverella* van Tieghem (1895) [Loranthaceae]. →*Echeveria.*

Oreosedum Grulich, Preslia 56, 41-42. 1984. Type: *Sedum album* Linnaeus. →*Sedum.*

Orostachys F.E.L. Fischer ex A. Berger, in Engler & Prantl, Nat. Pflanzenfam. ed. 2, 18a, 463. 1930. Type: *Umbilicus malacophyllus* Pallas (typification according to Borissova, Novost. Syst. Vyssh. Rast. 6, 119. 1969). Also attributed to Fischer (Cat. Jard. Gorenki, 99. 1808). Species: c. 10. **Dist.:** Inner Asia to Japan.

Notes: The genus appears to be monophyletic; it differs from *Sedum* primarily on account of the inflorescence architecture and the basally attenuate carpels. N.B.: The gender of this name is to be treated as feminine, along with *Stachys* and *Meterostachys* (see Moran, Cact. Succ. J. (US) 44, 262-273. 1972). A revision of the genus is in progress by H. Ohba and co-workers.

Pachyphytum Link, Klotzsch & Otto, Allg. Gartenzeit. 9, 9. 1841. Type: *Pachyphytum bracteosum* Klotzsch (typification by inference, only element included). Synonym: *Diotostemon* Salm-Dyck (1854). Species: c. 12. **Dist.:** Mexico.

Notes: Closely related to *Echeveria* (as testified by the x *Pachyveria* hybrids), but *Pachyphytum* can be hybridized many other North American genera (C.H. Uhl). All species here classified have been studied in detail by Moran (various publications in Cact. Succ. J. US).

Pagella Schönland, Ann. Bolus Herb. 3, 67, t. 3A. 1921. Type: *Pagella archeri* Schönland. Species: 1. **Dist.**: RSA (Cape Province).→*Crassula.*

Parvisedum Clausen, Cact. Succ. J. (US) 18, 58. 1946. Type: *Sedum pumilum* Bentham (typification by inference, with the type of the replaced name). Unnecessary *nom. nov. pro Sedella* Britton & Rose (1903) *non Sedella* Fourrier (1868). →*Sedella.*

Notes: Clausen thought that *Sedella* Britton & Rose was an illegitimate later homonym of *Sedella* Fourrier (1868), but the latter name is invalid, because Fourrier gave no description or diagnosis.

Perrierosedum (A. Berger) H. Ohba, J. Fac. Sci. Univ. Tokyo, Bot. 12, 166. 1978. Type: *Sedum madagascariense* Perrier. Based on *Sedum* sect. *Perrierosedum* A. Berger (in Engler & Prantl, Nat. Pflanzenfam. ed. 2, 18a, 446. 1930). Species: 1. **Dist.:** Madagascar.

Notes: A poorly-known relict element, probably related to *Hylotelephium* Ohba (l.c.), and not to *Cotyledon* (Fröderström, Acta Horti Gothob. 7, app., 97. 1932); or intermediate between *Hylotelephium* and *Cotyledon / Kalanchoe* (H. 't Hart).

Petrogeton Ecklon & Zeyher, Enum. Pl. Afr., 291, 1837. Type: *Petrogeton typicum* Ecklon & Zeyher. (according to typification of *Crassula* sect. *Petrogeton* by Tölken, Contrib. Bolus. Herb. 8, 217. 1977; Fl. South Afr. 14, 137. 1985). →*Crassula.*

Petrophyes Webb & Berthelot, Hist. Nat. Iles Canar. 3(2,1), 201. 1841. Type: not designated. An illegitimate (Art. 62.1) new name for *Monanthes.*→ *Monanthes.*

Petrosedum Grulich, Preslia, 56(1): 39, 1984. Type: *Sedum reflexum* Linnaeus. →*Sedum.*

Phedimus Rafinesque, Amer. Monthly Mag. & Crit. Review 1, 438. 1817. Lectotype: *Sedum stellatum* Linnaeus (designated by 't Hart, Fl. Medit. 1, 52. 1991). Usually attributed to Rafinesque in Catal., 15, 1824. →*Sedum.*

Notes: Suggested for treatment at generic rank by 't Hart (this volume), which would make *Asterosedum* a synonym of *Phedimus* rather than *Sedum.*

Physocalycium Vest, Flora 3, 409-410. 1820. Type: not designated. →*Bryophyllum.*

Pistorinia De Candolle, Prodr. 3, 399. 1828. Type: *Cotyledon hispanica* Linnaeus (typification by inference, only element included). Species: c. 4. **Dist.:** Iberian Peninsula, North Africa.

Notes: A little-known group with sympetalous flowers; cytologically they are not distinct from *Sedum* and can be interpreted as aberrant members of it.

Poenosedum Holub, Folia Geobot. Phytotax. 19, 98-99. 1984. Type: *Sedum tuberosum* Cosson & Letourneaux. Based on *Sedum* sect. *Rhodiola* subsect. *Primuloides* ser. *Tuberosae* A. Berger (in Engler & Prantl, Nat. Pflanzenfam. ed.2, 18a, 443. 1930). →*Sedum.*

Procrassula grisebach, Spicil. Fl. Rumel. 1:323,1843. Typus: *Crassula caespitosa* Cavanilles (= *Procrassula magnolii* Grisebach nom. illeg. Typification according to ING.) →*Sedum*

Prometheum (A. Berger) H. Ohba, J. Fac. Sci. Univ. Tokyo, Bot. 12, 168. 1978. Type: *Sedum sempervivoides* Fischer. This genus has priority by three months over *Pseudorosularia* Gurgenidze. Species: 2. **Dist.:** Anatolya, Northern Iran, Caucasus. → *Sedum*

Notes: In the original circumscription the genus comprised two species, but it can not be accepted because both species can be hybridized with *Sedum tymphaeum* (H. 't Hart). An alternative circumscription of the genus is proposed by 't Hart (this volume) in which *Prometheum* comprises *S. tymphaeum* and several species of *Rosularia* with a basic chromosome number x=7.

Pseudorosularia Gurgenidze, Zam. Sist. Geogr. Rast. 35, 10. 1978. Type: *Sedum pilosum* Bieberstein. *Prometheum* has priority over this name by three months. →*Sedum.*

Pseudosedum (Boissier) A. Berger, in Engler & Prantl, Nat. Pflanzenfam. ed. 2, 18a, 465. 1930. Type: *Umbilicus lievenii* Ledebour (typification by inference, only element included). Based on *Umbilicus* sect. *Pseudosedum* Boissier (Fl. Orient. 2, 775. 1872). Species: c. 10. **Dist.:** Caucasus, Inner Asia.

Notes: Nothing is known about the proper relationship of this group, which is morphologically similar to *Sedum.*

Purgosea Haworth, Phil. Mag. 3, 124. 1828. Type: not designated. An orthographic variant for *Turgosea* Haworth (1821). *Purgosia* G. Don (Gen. Hist. 3, 97, 103. 1834) is an orthographic variant. →*Crassula.*

Pyrgosea Sweet ex Ecklon & Zeyher, Enum., 298. 1837. Type: not designated. An orthographic variant for *Turgosea* Haworth (1821). →*Crassula.*

Rhodia Adanson, Fam. 2, 248. 1763. Type: not designated. A homotypic synonym of *Rhodiola* Linnaeus (Spec. Pl. 1753) according to ING. →*Rhodiola.*

Rhodiola Linnaeus, Spec. Pl., 1035. 1753. Type: *Rhodiola rosea* Linnaeus (Typification by inference, only element included). Synonyms: *Chamaerhodiola* Nakai (1934); *Clementsia* Rose (1903); *Kirpicznikovia* Löve & Löve (1976); *Rhodia* Adanson (1763); *Tolmachevia* Löve & Löve (1976). Species: c. 45. **Dist.:** North America, Europe, Inner & Eeastern Asia, Himalaya.

Notes: A highly derived group, differing from *Sedum* proper not only morphologically (forming a caudex with axillary, hapaxanth, and often deciduous floral stems) but also chemically (glycosides and flavonoids typical for *Rhodiola* are lacking in *Sedum*, see Kurkin & al., Rast. Resursy 1986, 310-

319. 1986). The same authors place *R. semenovii* in *Clementsia* on chemical evidence. Morphologically there are certain affinities between *Rhodiola* and *Hylotelephium.* Chloroplast DNA studies indicate that both genera have been derived from the same *Sedum* lineage, the so-called 'flat-leaved Asian sedums' (H. 't Hart).

Rhopalota N.E. Brown, Cact. Succ. J. (US) 3, 7. 1931. Type: *Crassula aphylla* Schönland & Baker (typification according to ING). →*Crassula.*

Rochea De Candolle, Hist. Pl. Succ., t. 103. 1802 (publ. 1810?). Type: *Crassula coccinea* Linnaeus (Conserved type). *Nomen conservandum in vice Rochea* Scopoli (1777) [Papilionaceae] (see ICBN, 106. 1952) *et Larochea* Persoon (1805). A. Berger gives the publication particulars as Prodr. 3, 393. 1828. →*Crassula.*

Rosularia (De Candolle) Stapf, Curtis's Bot. Mag. 149, t. 8985, in adnot. 1923. Lectotype: *Cotyledon sempervivum* M. Bieberstein (designated by A. Borissova, Novost. Syst. Vyssh. Rast. 6, 119. 1969.). Synonyms: *Afrovivella* A. Berger (1930); *Sempervivella* Stapf (1923). Species: 25. **Dist.:** Mediterranean, Northern & Eastern Africa, South-Western & Central Asia. **LST:** Eggli, U., Bradleya 6, suppl. 1988.

Notes: Without doubt closely related to *Sedum. Rosularia,* as circumscribed by Eggli (1988), is polyphyletic, and only the Asian taxa with a basic chromosome number x=9 (*R.* sect . *Rosularia* and some additional species) are maintained if the genus is split at the section level ('t Hart, this volume).

Sarcolipes Ecklon & Zeyher, Enum. Pl. Afr. 290. 1837. Type: *Sarcolipes pubescens* Ecklon & Zeyher (Typification by inference, only element included.). → *Crassula.*

Sedastrum Rose, North Amer. Fl. 22, 58. 1905. Type: *Sedum incertum* Hemsley. →*Sedum.*

Sedella Britton & Rose, Bull. New York Bot. Gard. 3(9), 45. 1903. Type: *Sedum pumilum* Bentham. *Non Sedella* Fourrier (1868) *nom. inval.* Synonym: *Parvisedum* Clausen (1946).Species: 4. **Dist.:** USA (Northern California). →*Sedum.*

Sedella Fourreau, Ann. Soc. Lyon, ser. nov. 16, 384. 1868. Lectotype: *Sedum atratum* Linnaeus (designated by 't Hart, Fl. Medit. 1, 52. 1991.). →*Sedum.*

Notes: Nom. inval. (Art. 32.1); taxa included were *Sedum atratum* and *S. villosum* (see Eggli, Bradleya 10, 83. 1992). *Non Sedella* Britton & Rose (1903).

Sedum Linnaeus, Spec. Pl., 430. 1753. Lectotype: *Sedum acre* Linnaeus (designated by Green in Hitchcock, Prop. Brit. Bot., 156. 1929; and proposed for conservation by 't Hart in Jarvis, Taxon 41, 569. 1992). This selection supersedes the choice of *Sedum telephium* Linnaeus by Britton & Rose (North Amer. Fl. 22. 1905) to maintain the familiar usage of the generic name when the segregate *Hylotelephium* is recognized. Synonyms: *Aithales* Webb & Berthelot (1840); *Aizopsis* Grulich (1984); *Amerosedum* Löve & Löve (1985); *Anacampseros* Miller (1754); *Asterosedum* Grulich (1984); *Breitungia* Löve & Löve (1985); *Cepaea* Fabricius (1759); *Chetyson* Rafinesque (1821); *Clausenellia* Löve & Löve (1985); *Cockerellia* (Clausen & Uhl) Löve & Löve (1985); *Congdonia* Jepson (1925); *Corynephyllum* Rose (1905); *Crassularia* Hochstetter ex Schweinfurth (1867); *Enchylus* Erhart (1789); *Etiosedum* Löve & Löve (1985);

Gormania Britton (1903); *Helladia* M. Král (1987); *Hjaltalinia* Löve & Löve (1985); *Leucosedum* Fourreau (1868); *Macrosepalum* Regel & Schmalhausen (1882); *Monanthella* A. Berger (1930); *Mucizonia* (De Candolle) A. Berger (1930); *Oreosedum* Grulich (1984); *Petrosedum* Grulich (1984); *Phedimus* Rafinesque (1817); *Poenosedum* Holub (1984); *Procrassula* Grisebach (1843); *Prometheum* (A. Berger) H. Ohba (1978); *Pseudorosularia* Gurgenidze (1978); *Sedastrum* Rose (1905); *Sedella* Britton & Rose (1903); *Sedella* Fourreau (1868); *Spathulata* (A. Borissova) Löve & Löve (1985); *Telmissa* Fenzl (1842); *Tetradium* Dulac (1867); *Tetrorum* Rose (1903); *Triactina* Hooker & Thomson (1858). Species: approx. 400. **Dist.:** Northern hemisphere; some in Central & Eeastern Africa. **LST:** Fröderström H., Acta Hort. Goth. 5, 6, 7, 10 (suppl.). 1930-35. (key to species, last world-wide revision); Clausen, R.T., *Sedum* of the Trans-Mexican Volcanic Belt. Comstock Publ., New York. 1959. (keys); Clausen, R.T., *Sedum* of North America North of the Mexican Plateau. Cornell Univ. Press, Ithaca. 1975. (keys); Hart, H. 't, Fl. Medit. 1, 31-61. 1991. (synopsis of European series, key).

Notes: The above estimate of species number is extremely tentative. The large number of synonyms illustrates the problems which have beset *Sedum* classification in recent years, and none of the recent divisions into segregates has met with approval. *Sedum* is still a rather heterogeneous and variable assemblage, even without the species now classified as *Hylotelephium*, *Orostachys*, and *Rhodiola*. The South American species probably all belong to the poorly understood *Villadia*. J. Meyrán (in litt, Jan. 1993) raises the interesting question whether sect. *Pachysedum* (with lateral inflorescences) should not be removed from *Sedum* and placed in the *Echeverioideae* (see under *Echeveria* for further comments).

Sedum Adanson, Fam. 2, 248. 1763. Type: not designated. *Non Sedum* Linnaeus (1753). A synonym of *Sempervivum* according to ING. → *Sempervivum.*

Sempervivella Stapf, Curtis's Bot. Mag. 1923, t. 8985, in adnot., 1923. Type: *Sempervivum album* Edgeworth (typification by inference, only element included). → *Rosularia.*

Sempervivum Linnaeus, Spec. Pl., 464. 1753. Lectotype: *Sempervivum tectorum* Linnaeus (according to Britton & Rose, North Amer. Fl. 22, 10. 1905). Synonym: *Sedum* Adanson (1763). Species: c. 25. **Dist.:** Europe to Iran, Caucasus, Morocco (Atlas).

Notes: See comments under *Jovibarba.*

Septas Linnaeus, Pl. Afr. Rar., 10. 1761 (dated 1760). Type: *Septas capensis* Linnaeus. *Non Septas* Loureiro (1790) [= *Heptas* Meisner (1840), Scrophulariaceae]. → *Crassula.*

Septimia P.V. Heath, Calyx 3, 104. 1993. Type: *Crassula spathulata* Thunberg. *Nom. nov. pro Crassula* group *Spathulatae* Schönland (Trans. Roy. Soc. South Africa 17, 164. 1929). → *Crassula.*

Sinocrassula A. Berger, in Engler & Prantl, Nat. Pflanzenfam. ed. 2, 18a, 462-463. 1930. Lectotype: *Crassula indica* Decaisne (designated by H. Ohba, J. Fac. Sci. Univ. Tokyo, ser. 3, 12, 170. 1978.). Species: 2-3. **Dist.:** Himalaya to South-Western China.

Spathulata (A. Borissova) Löve & Löve, Taxon 34, 163. 1985. Type: *Sedum spurium* S.G. Gmelin. Based on *Sedum* subsect. *Spathulata* A. Borissova (Novost. Sist. Vyss. Rast. 6, 117. 1969).→ *Sedum.*

Sphaeritis Ecklon & Zeyher, Enum. Pl. Afr., 299, 1837. Type: not designated. *Non Sphaeritis* Unger (1850) [fossil fungi]. → *Crassula.*

Stylophyllum Britton & Rose, Bull. New York Bot. Gard. 3(9), 33. 1903. Type: *Sedum edule* Nuttall ex Torrey & A. Gray. The type was originally given as *Cotyledon edulis* Nuttall, which is a later combination for *Sedum edule* Nuttall eventually published by Brewer (*Cotyledon edulis* (Nuttall) Brewer). →*Dudleya.*

Tacitus Moran, Cact. Suc. Mex. 19, 73, 75-84, 94. 1974. Type: *Tacitus bellus* Moran & Meyr†n. The name is frequently but erroneously attributed to Moran & Meyrán. Species: 1. **Dist.:** Western Mexico.

Notes: D.R. Hunt treated this monotypic genus as a synonym of *Graptopetalum*, but it does not share the rather uniform floral characteristics of *Graptopetalum.* It therefore seems to be impractical to unite the two, unless *Thompsonella* is also included (R. Moran, in litt. 1991). Moreover, *Graptopetalum* and *Tacitus* differ in the structure of the epidermal waxes (J. Thiede).

Telephium J. Hill, Brit. Herb., 36. 1756. Type: not designated. *Nom. illeg.* (Art. 64.1), *non Telephium* Linnaeus (1753) [Caryophyllaceae].→ *Hylotelephium.*

Telmissa Fenzl, Pugill. Pl. Nov. Syr., 14-15. 1842. Type: *Telmissa sedoides* Fenzl (typification according to ING). →*Sedum.*

Tetradium Dulac, Fl. Hautes-Pyren., 321. 1867. Type: *Rhodiola odorata* Lamarck. *Nom. illeg.* (Art. 64.1), *non Tetradium* Loureiro (1790) [Rutaceae]. →*Sedum.*

Tetraphyle Ecklon & Zeyher, Enum. Pl. Afr., 292. 1837. Type: not designated.→*Crassula.*

Tetrorum Rose, North Amer. Fl. 22, 59. 1905. Type: *Sedum pusillum* Michaux. Species: 1. **Dist.:** USA (granite outcrops). →*Sedum*

Notes: See Wilbur (Sida 13, 1-16. 1988) on the application of the name *Sedum pusillum* Michaux and nomenclature of *Diamorpha.*

Thisantha Ecklon & Zeyher, Enum. Pl. Afr., 302, 1837. Type: not designated.→*Crassula.*

Thompsonella Britton & Rose, Contr. US Nation. Herb. 12, 391. 1909. Type: *Echeveria minutiflora* Rose. Species: c. 3. Dist.: Southern Mexico. **LST:** Moran, R., Cact. Succ. J. (US) 41, 173-177. 1969; l.c. 64, 37-44. 1992.

Notes: Differs from *Graptopetalum* mainly in habit, inflorescence, and flower behaviour. The two could be united and placed in the synonymy of *Sedum* (H. 't Hart). Some similiarity with *Lenophyllum* is also noted.

Tillaea Linnaeus, Spec. Pl., 128, 1753. Lectoype: *Tillaea muscosa* Linnaeus (selected by inference by De Candolle, Bull. Sci. Soc. Philom. Paris 3, 2. 1801 [only taxon left]). P.V. Heath (Calyx 4, 84. 1994) attributes the selection of the lectotype to Britton & Rose (North Amer. Fl. 22. 1905). Synonyms: *Bulliarda* De Candolle (1801); *Combesia* A. Richard (1848); *Dasystemon* De Candolle (1828); *Helophytum* Ecklon & Zeyher (1836); *Mesanchum* Dulac (1867); *Tillaeastrum* Britton (1903). Species: approx. 20. **Dist.:** worldwide except Antarctica.

Notes: Schönland's (in Engler & Prantl, Nat. Pflanzenfam. 3, 2a. 1891) and Berger's (in Engler & Prantl, Nat. Pflanzenfam. ed. 2, 18a. 1930) classifications are generally accepted and nowadays *Tillaea* is only rarely upheld at generic level. However, this genus of specialized, tiny and often (semi-)aquatic species with a pronounced reduction of all parts, a different mode of opening of the follicles, and a small number of ovules certainly deserves to be reinstated. Nucleotide sequence analysis of chloroplast DNA indicates that genetic divergence between *Crassula* and *Tillaea* is of the same order of magnitude as among *Echeveria, Kalanchoe,* and *Sempervivum* (van Ham, this volume). It remains to be demonstrated that *Tillaea* in its traditional circumscription is indeed monophyletic, or whether some of the reductions might be the result of parallel developments.

Tillaeastrum Britton, Bull. New York Bot. Gard. 3(9), 1. 1903. Type: *Tillaea aquatica* Linnaeus. Illegitimate *nom. nov. pro Bulliarda* De Candolle (1801), *non Bulliarda* Necker (1790) *nom. inval.* →*Tillaea.*

Tillea Sanguinetti, Atti Accad. Pontif. Sci. Nuovi Linei 5, 506. 1852. Type: not designated. An orthographic variant for *Tillaea* Linnaeus. →*Tillaea.*

Toelkenia P.V. Heath, Calyx 3, 105. 1993. Type: *Cotyledon arborescens* Miller. *Nom. nov. pro Crassula* group *Arborescens* Schönland (Trans. Roy. Soc. South Africa 17, 165. 1929). →*Crassula.*

Tolmachevia Löve & Löve, Bot. Not. 128: 514, 1976. Type: *Rhodiola integrifolia* Rafinesque. Species: 3. **Dist.:** Wwester North America, Eeastern Asia. → *Rhodiola*

Notes: Said to differ from *Rhodiola, Kirpicznikovia,* and *Clementsia* primarily by a different base number and the polygamous flowers.

Triactina Hooker & Thomson, J. Linn. Soc. 2, 90 [103 ?]. 1858. Type: *Triactina verticillata* Hooker & Thomson (= *Sedum triactina* A. Berger; typification by inference, only element included). →*Sedum.*

Turgosea Haworth, Rev. Pl. Succ., 14. 1821. Type: not designated. →*Crassula.*

Tylecodon Tölken, Bothalia 12, 378. 1978. Type: *Cotyledon cacalioides* Linnaeus fil. Species: c. 35. **Dist.:** RSA (Cape Province), Southern Namibia. **LST:** Tölken, H.R., Bothalia 12, 378-382. 1978; Fl. South Afr. 14, 19-35. 1985. (key).

Umbilicus De Candolle, Bull. Sci. Soc. Philom. Paris 3, 1. 1801. Type: *Cotyledon umbilicus* Linnaeus (typification according to ING). Often erroneously attributed to De Candolle (Prodr. 3, 400. 1828). Synonyms: *Chiastophyllum* (Ledebour) Stapf ex A. Berger (1930); *Cotyliphyllum* Link (1831). Species: c. 12. **Dist.:** Macaronesia, Mediterranean, Near East, Arabia, North & (North-)East Africa.

Notes: Chiastophyllum (1 sp. from the Western Caucasus) is cytologically identical to *Umbilicus* and differs primarily in its non-peltate opposite leaves.

Urbinia Britton & Rose, Bull. New York Bot. Gard. 3(9), 11. 1903. Type: *Echeveria agavoides* Lemaire. →*Echeveria.*

Vauanthes Haworth, Rev. Pl. Succ., 18. 1821. Type: *Crassula dichotoma* Linnaeus (= *Vauanthes*

chloraefolia Haworth, *nom. illeg.*; typification according to ING). →*Crassula.*
Notes: The spelling is given as *Vausanthes* by P.V. Heath (Calyx 3, 113. 1993).

Verea Willdenow, Spec. Pl. 2, 471. 1799. Type: not designated. An orthographic variant for *Vereia.* →*Kalanchoe.*

Vereia H. Andrews, Bot. Repos. 1, t. 21. 1798 [dated 1797 ?]. Type: *Vereia crenata* H. Andrews (typification according to ING). →*Kalanchoe.*

Villadia Rose, Bull. New York Bot. Gard. 3(9), 3. 1903. Type: *Cotyledon parviflora* Hemsley (*non Cotyledon parviflora* Desfontaines). Synonym: *Altamiranoa* Rose (1903). Species: c. 40. **Dist.:** USA (Texas), Mexico, Central & South America.

Notes: The relationships of this genus versus *Sedum* need proper evaluation; the character of partially fused petals is very weakly developed in several taxa, and complete intergradation with *Sedum* seems to be present. Monophyly cannot be definitely assumed (M. Kimnach). Moreover, hybrids with *Sedum* seem to be possible. Epidermal wax structures of at least some taxa of *Villadia* differ from those found in *Sedum* (J. Thiede).